T0214180

Digital Da Vinci

Newton Lee
Editor

Digital Da Vinci

Computers in Music

 Springer

Editor
Newton Lee
Newton Lee Laboratories, LLC
Tujunga
California
USA

ISBN 978-1-4939-5583-1   ISBN 978-1-4939-0536-2 (eBook)
DOI 10.1007/978-1-4939-0536-2
Springer New York Heidelberg Dordrecht London

Printed on acid-free paper

Springer is part of Springer Science+Business Media (www.springer.com)

"Do you know that our soul is composed of harmony?"

—Leonardo da Vinci

About the Book

The *Digital Da Vinci* book series opens with the interviews of music mogul Quincy Jones, MP3 inventor Karlheinz Brandenburg, Tommy Boy founder Tom Silverman, and entertainment attorney Jay L. Cooper. A strong supporter of science, technology, engineering, and mathematics programs in schools, The Black Eyed Peas founding member will.i.am announced in July 2013 his plan to study computer science.

Leonardo da Vinci, the epitome of a Renaissance man, was an Italian polymath at the turn of the 16th century. Since the Industrial Revolution in the 18th century, the division of labor has brought forth specialization in the workforce and university curriculums. The endangered species of polymaths is facing extinction. Computer science has come to the rescue by enabling practitioners to accomplish more than ever in the field of music.

In this book, Newton Lee recounts his journey in executive producing a Billboard-charting song like managing agile software development; M. Nyssim Lefford expounds producing and its effect on vocal recordings; Dennis Reidsma, Mustafa Radha, and Anton Nijholt survey the field of mediated musical interaction and musical expression; Isaac Schankler, Elaine Chew, and Alexandre François describe improvising with digital auto-scaffolding; Shlomo Dubnov and Greg Surges explain the use of musical algorithms in machine listening and composition; Juan Pablo Bello discusses machine listening of music; Stephen and Tim Barrass make smart things growl, purr, and sing; Raffaella Folgieri, Mattia Bergomi, and Simone Castellani examine EEG-based brain-computer interface for emotional involvement in games through music; and last but not least, Kai Ton Chau concludes the book with computer and music pedagogy.

Digital Da Vinci: Computers in Music is dedicated to polymathic education and interdisciplinary studies in the digital age empowered by computer science. Educators and researchers ought to encourage the new generation of scholars to become as well rounded as a Renaissance man or woman.

Contents

Contributors

Stephen Barrass Faculty of Arts and Design, University of Canberra, Canberra, Australia

Tim Barrass Independent Artist and Inventor, Melbourne, Australia

Juan Pablo Bello Music and Audio Research Laboratory (MARL), New York University, New York, USA

Mattia G. Bergomi Dipartimento di Informatica, Università degli Studi di Milano, Milano, Italy

Ircam, Université Pierre et Marie Curie, Paris, France

Simone Castellani CdL Informatica per la Comunicazione, Università degli Studi di Milano, Milano, Italy

Kai Ton Chau Kuyper College, Michigan, USA

Elaine Chew Queen Mary University of London, London, UK

Shlomo Dubnov Music Department, University of California in San Diego, San Diego, USA

Raffaella Folgieri DEMM, Dipartimento di Economia, Management e Metodi quantitativi, Università degli Studi di Milano, Milano, Italy

Alexandre R. J. François Interactions Intelligence, London, UK

Newton Lee Newton Lee Laboratories, LLC, Tujunga, CA, USA

School of Media, Culture & Design, Woodbury University, Burbank, CA, USA

M. Nyssim Lefford Luleå University of Technology, Luleå, Sweden

Anton Nijholt Human Media Interaction, University of Twente, AE, Enschede, The Netherlands

Mustafa Radha Human Media Interaction, University of Twente, AE, Enschede, The Netherlands

Dennis Reidsma   Human Media Interaction, University of Twente, AE, Enschede, The Netherlands

Isaac Schankler   Process Pool Music, Los Angeles, CA, USA

Greg Surges   Music Department, University of California in San Diego, San Diego, USA

About the Authors

Stephen Barrass is a researcher and academic at the University of Canberra where he lectures in Digital Design and Media Arts in the Faculty of Arts and Design. He holds a B.E. in Electrical Engineering from the University of New South Wales (1986) and a Ph.D. titled Auditory Information Design from the Australian National University (1997). He was a Post-Doctoral Fellow at the Fraunhofer Institute for Media Kommunication in Bonn (1998) and Guest Researcher in Sound Design and Perception at IRCAM in Paris (2009).

Tim Barrass has a background in electronic arts practice spanning over 20 years. In his visual and sound work he has explored ways of generating and understanding patterns of interaction in complex systems. He spent many years as a circus musician, developing custom software for electroacoustic instrumental performance in unpredictable circumstances. In recent times his focus has been on developing Mozzi, the sound synthesis library for Arduino. He is currently researching the ergonomics of typing with a cockatiel on each forearm.

Juan Pablo Bello is Associate Professor of Music Technology at New York University, with courtesy appointments at the Department of Electrical and Computer Engineering, and NYU's Center for Data Science. In 1998 he received a BEng in Electronics from the Universidad Simón Bolívar in Caracas, Venezuela, and in 2003 he earned a doctorate in Electronic Engineering at Queen Mary, University of London. Juan's expertise is in digital signal processing, computer audition and music information retrieval, topics in which he actively teaches, researches and publishes. His work has been supported by public and private institutions in Venezuela, the UK, and the US, including a CAREER award from the National Science Foundation. He co-founded the Music and Audio Research Lab (MARL), where he leads research on music informatics.

Mattia G. Bergomi is a mathematician, Ph.D. student in Computer Science, and member of the Laboratory of Music and Computer Science (LIM). His research interest lies in the intersection between music and mathematics: On one side the representation of musical objects with instruments borrowed from the Algebraic Topology; on the other side the new analysis methods given by the Computational Algebraic Topology and their interaction with machine learning algorithms.

Simone Castellani is a student in Computer Science at Università degli Studi di Milano. In his thesis he developed experiments in quantitative and qualitative analysis of perception of the emotional interaction between visual and audio stimuli. His research interests are the analysis of the brain responses to multilayers stimuli and its application in artificial intelligence.

Kai Ton Chau is Associate Professor and the Jack Van Laar Endowed Chair of Music and Worship at Kuyper College in Grand Rapids, Michigan. He directs the college choir and ensemble, teaches several music courses, and chairs the Arts and Sciences department. His diverse career in Hong Kong, Canada, and the United States has afforded him the opportunities to serve at various churches, inter-church events, mass choirs, and institutions of higher education (including Institute for Christian Studies in Toronto, and Redeemer University College in Ancaster, Canada). Chau earned an Honors diploma in composition at the Hong Kong Baptist University, a Master of Music in choral conducting at the University of Missouri-Kansas City, an MBA from Laurentian University in Ontario, Canada, and a doctorate in worship studies from the Robert E. Webber Institute for Worship Studies in Orange Park, Florida. He also holds professional designations (CGA, FCCA, CFP) in accounting and financial planning from Canada and the U.K.

Elaine Chew is Professor of Digital Media at Queen Mary University of London. A pianist and operations researcher by training, her research centers on the mathematical and computational modeling of aspects of performance, including music prosody, cognition, structure, and interaction, so as to make explicit what it is that musicians do, how they do it, and why. Previously, she was an Assistant then Associate Professor at the University of Southern California, where she founded the Music Computation and Cognition Laboratory; and, she held visiting positions at Harvard University and Lehigh University. Her research has been recognized by the US National Science Foundation Faculty Early Career Development Award and the Presidential Early Career Award in Science and Engineering, and a fellowship cluster on Analytical Listening through Interactive Visualization at the Radcliffe Institute for Advanced Study. She received PhD and SM degrees in Operations Research at MIT, a BAS in Mathematical and Computational Sciences (honors) and in Music (distinction) at Stanford, and FTCL and LTCL diplomas in Piano Performance from Trinity College, London.

Shlomo Dubnov is director of the Center for Research in Entertainment and Learning at University of California San Diego. He teaches in the Music and Interdisciplinary Computing in the Arts programs. Previously, he was a researcher at the Institute for Research and Coordination of Acoustics and Music (IRCAM) in Paris, and head of the multimedia track for the Department of Communication Systems Engineering at Ben-Gurion University. He is a senior member of IEEE and secretary of IEEE's Technical Committee on Computer Generated Music. He graduated from the Jerusalem Music Academy in composition and holds a doctorate in computer science from the Hebrew University, Jerusalem.

Raffaella Folgieri PhD in Computer Science, is Assistant Professor in Computer Skills at the Faculty of Political Science and of Information Technology at the Faculty of Political Science and at the Faculty of Medicine (Medical and Pharmaceutical Biotechnologies). She also teaches Information Technology Representation of Knowledge in the post-degree course in Cognitive Science and Decision Making, Virtual Reality in the Information Technology and Digital Communication degree course and Project Management at the Faculty of Mathematics, Physics, and Natural Sciences of the University of Milan. A member of the Italian Society of Engineering and of SIREN (Italian Neural Networks Society), she has published her research in several journal articles (main fields of interests: Brainomics; Brain Computer Interfaces; Virtual Reality; Bioinformatics; Machine Learning and AI; Quality assessment in complex software development; e-learning). Her work explores some of the central issues in cognitive research such as how people move from skilled performance to problem solving, how a person learns, manages errors, interprets visual stimuli, and communicates. She coordinates the research group Beside, focused on interpersonal, machine-machine and brain-machine communication mediated by technology, and ExCog (jointly with Prof. Lucchiari), aiming to study MIND in all its complexity and all possible shapes.

Alexandre R. J. Francois is a software engineer and an independent researcher, whose work has focused on the modeling and design of interactive (software) systems, as an enabling step towards the understanding of perception and cognition. His interdisciplinary research projects explored interactions within and across music, vision, visualization and video games. He was a 2007-2008 Fellow of the Radcliffe Institute for Advanced Study at Harvard University; a Visiting Associate Professor at Harvey Mudd College; a Visiting Assistant Professor at Tufts University; and, a Research Assistant Professor at the USC Viterbi School of Engineering. Prior to that, he was a Research Associate with the USC Integrated Media Systems Center and Institute for Robotics and Intelligent Systems. He holds PhD and MS degrees in Computer Science from USC, the Diplôme d'Etudes Approfondies (MS) from the University Paris IX - Dauphine (France), and the Diplôme d'Ingénieur from the Institut National Agronomique Paris-Grignon (France).

Newton Lee is founding director of the Woodbury University Digital Media Lab and adjunct professor of Media Technology at the School of Media, Culture & Design. He is also CEO of Newton Lee Laboratories LLC, president of the Institute for Education, Research, and Scholarships, and founding editor-in-chief of ACM Computers in Entertainment. Previously, he was a research scientist at AT&T Bell Laboratories, senior producer and engineer at The Walt Disney Company, research staff member at the Institute for Defense Analyses, and research scientist at Virginia Tech Library Systems. Lee graduated Summa Cum Laude from Virginia Tech with a B.S. and M.S. degree in Computer Science, and he earned a perfect GPA from Vincennes University with an A.S. degree in Electrical Engineering and an honorary doctorate in Computer Science. He is the co-author of *Disney Stories: Getting to Digital;* the author of the Total Information Awareness book series including

Facebook Nation and *Counterterrorism and Cybersecurity;* and the editor of the Digital Da Vinci book series including *Computers in Music* and *Computers in the Arts and Sciences.*

M. Nyssim Lefford   is visiting senior lecturer at Luleå University of Technology in Sweden. Previously, she was visiting assistant professor at Georgia Institute of Technology. Her investigations have ranged from record production to installation art to music cognition research. She has sought to understand how creators create, to find ways to facilitate the creative process, and to uncover the potential in new technology and new perspectives to shape future creations. She received her Ph.D. and M.S. from the MIT Media Lab, and her Bachelor of Music in music production and engineering and film scoring from Berklee College of Music.

Anton Nijholt   started his professional life as a programmer at TNO-Delft. He studied civil engineering, mathematics and computer science at Delft University of Technology and did his Ph.D. in theoretical computer science at the Vrije Universiteit in Amsterdam. He held positions at the University of Twente, the University of Nijmegen, McMaster University (Canada), the Vrije Universiteit Brussels (Belgium), and at NIAS in Wassenaar. During some years he was scientific advisor of Philips Research Europe. Presently he is member of the Human Media Interaction group of the University of Twente. His main research interests are multi-party interaction, multimodal interaction, brain-computer interfacing and entertainment computing.

Mustafa Radha   is a graduate student in Human-Media Interaction at the University of Twente. He is mainly interested in the social and computational semiotics of augmented human life.

Dennis Reidsma   is Assistant Professor at the Human Media Interaction group and Lecturer of the Creative Technology curriculum at the University of Twente. After receiving his MSc degree in Computer Science cum laude for a thesis on semantic language processing, Dennis Reidsma completed his PhD degree at the Human Media Interaction group of the University of Twente. He supervises a number of BSc, MSc, and PhD students on topics of computational entertainment and interactive playgrounds, runs several research projects in this area, and is regularly involved in the organization of conferences such as INTETAIN and ACE. In addition, he has published many papers on interaction with Virtual Humans, and consolidated the results of this joint work with Herwin van Welbergen in the release of Elckerlyc, a state-of-the-art Open Source software platform for generating continuous interaction with Virtual Humans.

Isaac Schankler   is a Los Angeles-based composer and improviser. He co-directs the concert series People Inside Electronics and teaches music technology at the University of Southern California (USC). He has published in the International Journal of Arts and Technology, Computer Music Journal, and the proceedings of international conferences including MCM (Mathematics and Computation in Music), ISPS (International Symposium of Performance Science), and ICME

(International Conference on Multimedia & Expo). He also writes a regular column at NewMusicBox, the online publication of New Music USA. He completed his doctoral studies in composition at the USC Thornton School of Music, and holds degrees in composition (MM, BM) and English (BA) from the University of Michigan.

Greg Surges makes electronic music, software, and hardware. His research and music have been presented at SMAC/SMC, the International Computer Music Conference, NIME, the SPARK Festival in Minneapolis, and the SEAMUS National Conference. He is currently a PhD student at the University of California, San Diego. Previously, he earned a MM in Music Composition and a BFA in Music Composition and Technology at the University of Wisconsin – Milwaukee. He currently lives in San Diego, CA with his wife and cat.

Chapter 1
A Tale of Four Moguls: Interviews with Quincy Jones, Karlheinz Brandenburg, Tom Silverman, and Jay L. Cooper

Newton Lee

MP3 and peer-to-peer file sharing technology single-handedly disrupted the age-old music business. iTunes and YouTube have displaced record stores and MTV. If we take the cue from Netflix which has successfully produced original content, it will not be long before Apple and Google will sign new artists and rival the record labels.

1.1 Interview with Quincy Jones

Quincy Jones, who composed more than 40 major motion picture and television scores, has earned international acclaim as producer of the best-selling album of all time—Michael Jackson's *Thriller*—which has sold more than 110 million copies worldwide. Jones was also the producer and conductor of the charity song "We Are the World."

The all-time most nominated Grammy artist with a total of 79 nominations and 27 wins, Jones has also received an Emmy Award, 7 Oscar nominations, the Academy of Motion Picture Arts and Sciences' Jean Hersholt Humanitarian Award, the Ahmet Ertegun Award for Lifetime Achievement, and the Grammy Legend Award. He was inducted into the Rock & Roll Hall of Fame in 2013.

On November 25, 2003, I had the honor to interview Quincy Jones at his then residence at 1100 Bel Air Place. With the assistance of my colleagues Eric Huff, Brett Hardin, and Laura Knight, we took our video cameras and lighting equipment to the three-story wood-and-brick house, checked in with the security guards, and met with a big warm welcome from Quincy Jones (see Fig. 1.1).

The following is a transcript of the video interview with Quincy Jones (Lee 2004a):

N. Lee (✉)
Newton Lee Laboratories, LLC, Tujunga, CA, USA
e-mail: newton@newtonlee.com

School of Media, Culture & Design, Woodbury University, Burbank, CA, USA
e-mail: newton.lee@woodbury.edu

N. Lee (ed.), *Digital Da Vinci,* DOI 10.1007/978-1-4939-0536-2_1,
© Springer Science+Business Media New York 2014

1

Fig. 1.1 Quincy Jones interviewed by Newton Lee on November 25, 2003

Newton Lee What do you think about educating children through entertainment?

Quincy Jones Entertainment, I think, is the best way, that's my first reaction to [programming language] Squeak, with the kids that are having fun while they're learning. I think that's the best way. It shouldn't be some amazing obstacle course that you got to master. It can be fun. I think music—anything kind of knowledge—takes repetition. There's an old expression: Whatever you wish to do with these, you should first learn to do through diligence. You know, you have to. It's like with music: It's scales, it's diligence, a lot of discipline which makes your life easier later on, and if kids can have fun, why make it hard, you know? They're in alpha state until they're about five or six years old—I've got seven kids, I understand children a lot. If they're not broke, they don't need to be fixed. It's just a question of just letting them become familiar with information in a way that feels natural, so they enjoy it. They absorb more. If we retain 10 % of what we hear and 30 % of what we see and 80 % of what we do, then you can throw all the things in, and it can be a fun process. I mean, that's what we wish it to be anyway.

Lee What possible effect do you think music has on children?

Jones Oh, music has a powerful effect on everybody: Children, adults, animals, plants, everything. It's, I think, one of the abstract miracles and phenomena. I'm glad I made up my mind to be in it a long time ago because you can't see it, you can't smell it, you can't touch it, you can't taste it, but you can feel it. It can just light up your whole soul. And I think it can have a physical effect on your soul, in terms of strength and belief system, and so forth.[1]

Lee How has technology affected the way you create and produce music?

---

[1] Author's note: Three months before 16-year-old Olivia Wise died of brain tumor, she recorded her version of Katy Perry's "Roar" in September 2013 (Ganim and Stapleton 2013). Unable to stand and struggling with her breaths, she smiled as she sang "I got the eye of a tiger, a fighter, dancing through the fire" (Wise 2013). Her song on YouTube and iTunes has helped raise funds in support of brain tumor research.

Jones Well, technology has had an amazing effect on my life. I was fortunate enough to run into, I guess at the age of 13, 14, down in Seaside, Oregon, a man who is noted to be one of the pioneers of stereophonic sound. In 1948 he put earphones on my head in a ballroom, and it had two different signals, and I never heard that sound again, until it was live and we were recording Ray Charles in 1958 in New York City. Ray and I were kids during that time in Seattle, 14 and 16, we didn't really pay too much attention to technology then. But during the course of, oh my God, since from a very early, early age, I just happened to be at the spot where Leo Fender brings the Fender Bass to Monk Montgomery. We didn't know what it was. We were doing a lot of jazz dates; we were on our way to Europe. And, it turns out, I guess 1999, I did an interview for *USA Today*, and on the cover they had Steve Case, and another lady, a chief executive, and Bill Gates; and they were asking each one of us in each particular fields what singular piece of technology had the most effect on your particular genre, and I said, at that time, the Fender bass, because without the Fender bass which locked in with the electric guitar, which came from 1939, without the Fender bass there'd be no Rock & Roll, and there'd be no Motown, and a lot of things. It affected the way the notes were played.

And I guess in 1964, I used the first synthesizer when I wrote the theme for *Ironside* for Raymond Burr, and, you know, from the very beginning, we've started with 78 mono discs and we've gone through the whole cycle all the way through digital—that and binary numbers—it's phenomenal. We can really change the silicon technology to microchips to nanotechnology and the beat goes on; it doesn't stop. And I love it; I love the idea of it. I bet back in the day when Bach was alive that he wishes he could run into a synthesizer.

We were guinea pigs for a lot of the synthesizers in California because we wrote the movie scores. People like David Page, Lionel Richie, and Herbie Hancock—we were all guinea pigs when everything came out from YC30 to YC40 to YC80 to the algorithmic technology and everything else. It's been kind of fun to see how it progresses. I remember being the Vice President of Phillips in 1962. We went to Eindhoven Holland, and they had an experimental laboratory there with about like 8,000 scientists there. And they showed us in 1962 the first audio cassette and the first dime-thin laser beam disc. And I think it was the CFO, a man named Uttermelon that said, "It'll take us 35 years to really launch this." And when I look back, and it's very true because I saw it launched by RCA-Victor, and it never got to the finish line, and also by MCA when they had DiscoVision, I thought they would try to tie it in with the disco market. And in 1994, I think in June, I was in Tokyo on a promotional tour, I was invited to go to Toshiba the next morning to meet the president, Mr. Sato, and also see the demonstration of DVD which was A and B with the video disc, the same video disc we saw in 1962, and the DVD actually crushed the video disc into dust because the capacity is so far greater. I don't know. I enjoy the ride. I think it's great. I think the more we get into technology, people like Alan Kay and other people constantly pushed us into a more spiritual state, more with the spiritual. Keep the emotion lotion flowing! So, I think it's all good you know.

Even from the motion picture side, I remember a great friend of mine, the late Hal Ashby, that's his chair over there, he was a great friend, he called me one morning at I guess 5:30 in the morning, and said, "You've got to come down real quick, and see this new camera." It was a steady cam, and so snorkels and everything, you know, it all facilitates us to help us be more creative and you get inspired by what it can do, and it gives you another outlook. And I guess it worked both ways too because the things like the Linn 9000 Drum Machine can make drummers lazy, so they don't have to learn how to play good first, but it's in the eye of the beholder really.

Lee What do you think about digital music distribution over the Internet as a new business model?

Jones Well, I've been waiting for it a long time. The Internet is certainly the way now, even before that with the Baby Bells. I remember we were in talks with Bell Atlantic and Bell South, you know, trying to deal with fiber optic and twisted pair. At one time I think it would have been a big boom and a great platform for digital distribution, record distribution, because when you get into coaxial cable and it gets to be impulse buying, it's so easy to have a database on people's likes and dislikes with music, I think it would have been very good for the music business, to pull it up. But it'll get there anywhere. You know, Sean Fanning is a very good friend of mine, we've been hanging out lately, and he, at 18 years old, he's made it very clear there's a different direction with Napster, and now he's on the other side, he's doing some things like acoustic fingerprint, and so forth. He's experimenting with a lot of wonderful things.

1.2   Interview with Karlheinz Brandenburg

Experimenting with digital audio coding and perceptual measurement techniques, doctoral student Karlheinz Brandenburg wrote his dissertation in 1989 that became the foundation of MPEG-1 Layer 3 (MP3) and MPEG-2 Advanced Audio Coding (AAC). Right when I left AT&T Bell Laboratories in 1989, Brandenburg joined the same research labs in New Jersey and worked on perfecting the mp3 algorithm. In 1990, he returned to the University of Erlangen-Nuremberg in Bavaria, Germany. Since 2004, he has been the director of the Fraunhofer Institute for Digital Media Technology in Ilmenau. For his important role in the development of MP3, Brandenburg was inducted into the Consumer Electronics Association (CEA)'s CE Hall of Fame in 2007 along with other MP3 co-developers Heinz Gerhäuser and Dieter Seitzer (Consumer Electronics Association 2013).

During Brandenburg's visit to Los Angeles on July 22, 2004, I had a wonderful opportunity to ask him about MP3. The following is a transcript of the video interview with Karlheinz Brandenburg (Lee 2004b):

Newton Lee How did you create the MP3 algorithm by analyzing how the human ear and brain perceive sound?

Karlheinz Brandenburg Oh, that was a long time ago, and in fact is a long story starting with my thesis advisor who at some point had the idea: "What's the equivalence of bringing color to television in the audio area?" So his idea was to bring music over the telephone lines, and in fact he applied for a patent, and the patent examiner said, "No, it's not possible." So he looked for some graduate students to show it's possible anyway, and that started the research that led to some fresh ideas looking at site acoustics, the science of how our ears work and our brains work to perceive sound, and lots of signal processing involved. And in fact, it was not just myself, but a lot of people in the team in Erlangen, my colleagues in AT&T Bell Labs, and so on—working together as a standardization group to do the best audio coding possible at that time.

Lee So you had worked nearly 20 years, to develop and bring to markets the MP3 format, did you ever foresee the huge popularity of MP3 music?

Brandenburg I have to say yes and no. I remember some fifteen years ago, somebody asked me: "What will happen with this technology?" My answer was, "Don't know for sure, it could end up just in the library, like so many other theses, or it could be an international standard used by millions of people." I didn't dream of hundreds of millions of people and in fact five years later, some ten years ago, MP3 was there but apart from some professional applications, there was not much use for it. And we used the Internet to spread the news for marketing and we thought of the Internet as an application area. And I remember some meeting where in the team at Erlangen at the Fraunhofer Institute, where we said, "Oh, we have a window of opportunity to make MP3 the Internet audio standard. Of course, we didn't fully think what that would mean."

Lee Your current research is the spatial audio technology called Wave Field Synthesis. Can you tell us what it is?

Brandenburg That's based on basic research of Delft Technical University in the Netherlands. The idea is to recreate waveforms by superimposing the results of a lot of simple, small loud speakers. So you have a ring of loudspeakers around your room, and feed every loudspeaker with exactly the right signal and phase and amplitude and filter. The superposition of these wave fields generates a complete system of waveforms in the room like in reality. And that gives you a sense of spatialization, of immersion into a sound field, which is an order of magnitude better than everything we are used to. In 5.1, we always have the problem of the sweet spot and that if we turn our heads, the waveforms are not the right ones—they are different from reality. With IOSONO, which is our typemark name for Wave Field Synthesis, it's much more realistic, you can really have bullets flying through the room or a ghost whispering into your ears.

Lee What other new technologies are you working on?

Brandenburg In the Fraunhofer Institute in Ilmenau, we are working on a number of different techniques, some in the so-called "metadata" area. For example, technology to recognize music: There are already services that do that, like if you hear some music that you like it but you don't know what it is or where to get it, you can take your cell phone, dial some number, hold the cell phone in the direction to the music and after ten to fifteen seconds, the system will hang up and send a short message back to you telling you what music it is and in fact where to buy it, or just say, "Reply if you want to buy it."

Another technique is called "Query by Humming." That's again if you remember some music, or you're in a karaoke situation, and you want to find out what it is, you can sing, hum or whatever into the computer. If your singing is somewhat near the real tones, then it will answer back saying, "Oh yes! This fragment of music turns up in this song."

We have other technologies still in the work. For example: Audio-coding together with people in Erlangen on MPEG-4, on lossless audio coding, on the so-called MP3 Surround in Erlangen—a very nice advanced technique for compatible 5.1 with very low data rates, and for communication purposes there's very low delay audio-coding as well.

Another area of interest is digital rights management. I think there must be better compromises than today. We have today a number of different systems. I think Apple does it well with its system, but still Apple plays only on Apple devices. In the end there should be a standard. We are working on different basic techniques for digital rights management, including one we call "Lightweight Digital Rights Management" which does it sometimes the other way around. Normally, with digital rights management you are not allowed to do something, and it will just not work. That makes it very difficult to use some music which you've paid for in your home environment on different computers and so on. With our system, if you agree to include your electronic signature into the music, you can use it at different places. If you do that something wrong like putting it up on Kazaa, then somebody can find it, analyze it, find out you are the bad guy, and go after you.

1.3   Interview with Tom Silverman

MP3 and peer-to-peer file sharing technology single-handedly disrupted the age-old music business. iTunes and YouTube have displaced record stores and MTV. If we take the cue from Netflix which has successfully produced original content, it will not be long before Apple and Google will sign new artists and rival the record labels.

On February 15, 2011 at the New Music Seminar in Los Angeles, I met Tom Silverman, founder and CEO of Tommy Boy and co-founder of the New Music Seminar. Artists who have recorded for Tommy Boy Records include Coldcut, Queen Latifah, RuPaul, and Kristine W. (see Fig. 1.2).

Silverman serves on the boards of SoundExchange, Recording Industry Association of America (RIAA), American Association of Independent Music (A2IM), and Merlin Network. He received the National Academy of Recording Arts and Sciences Heroes Award in 2000 and the second annual Libera Awards Lifetime Achievement Award in 2013. The following is a transcript of the video interview with Tom Silverman (Lee 2009):

Fig. 1.2 Tom Silverman interviewed by Inessa Lee on February 15, 2011

Inessa Lee What is the New Music Seminar about?

Tom Silverman The New Music Seminar is the place where the visionaries come together to build a new music business. That's the purpose of it, and when artists come to find new ways to define success, and to bring their career to a whole new level by using totally different methods of thinking and acting.

Lee How has computer technology changed music?

Silverman Computer technology has totally changed the way people interact with music. Computer technology is everything from cell phones to computers themselves, and there are chips in almost everything right now. And nothing has been more affected than music by computer technology: The way music is made, the recording process, the way it's distributed, making it ubiquitous and easy to get music distributed. Everything is different now with music. And also, the way people consume music, they're listening on their telephones now to music, and they're listening on their iPods to music and their iPads to music. People will be listening to music in the future, 50 to 75 % of the time on devices that have a screen built into it, and that's a gigantic change! So how people integrate with music has totally been reformed because of computer technology.

Lee How do you see the music industry evolves in the future?

Silverman The music industry is going to change to be a business that's less product orientated, less based on records, and more based on relationship. We'll use songs to break artists, and to build brands for artists like we always have, but then we'll take those songs and those brands and the relationships between those artists and their fans, or the music and the fans of that music, and we'll find new ways of monetizing that relationship. So it becomes a relationship business as opposed to a product business. We will still sell products or will license products, but we'll also license music in different ways: We'll be involving with touring, we'll be involved with publishing, we'll be involved with merchandising, we'll be involved with restaurants, and we'll be involved with Dre's Beat headphones—anything no matter what your artist is. There's no limit to what you can do because the artist becomes a brand that stands for something. Lady Gaga stands for this, Dr. Dre stands for this, and Susan Boyle stands for this. Every artist stands for something. So there are so many artists, and they more they stand for, the more valuable they will be and there's more things they can do. They can be involved with products like 50 Cent was given a piece of Vitamin Water and then Vitamin Mineral Water sold for how many billions of dollars and he got his share of that. He made more money selling his share of Vitamin Water than he ever made in the music business, so it's not necessarily about music. I started Queen Latifah's career when she was 16, we never even sold half a million albums, but now she's one of the biggest stars in Hollywood. So you don't know where the business is going to go.

The new business will be about this: We're going to make a relationship with you, we're going to create a company around you, and everything you do goes into the company, and we're going to invest in you. It might be records, it might be acting, it might be books, it might be music publishing and songwriting, it could even be a chain of restaurants at the end of the day, but it's your brand that we're building.

1.4   Interview with Jay L. Cooper

With a client list that includes Katy Perry, Jerry Seinfeld, Mel Brooks, Sheryl Crow, and I, Jay L. Cooper is one of the most prominent entertainment attorneys in the United States. A shareholder at the international law firm Greenberg Traurig LLP, Cooper's law practice focuses on music industry, motion picture, television, multimedia, and intellectual property issues. He represents individuals and companies on intellectual property matters including recording and publishing agreements for artists and composers; actor, director, producer, and writer agreements in film and television; executive employment agreements; complex acquisitions and sales of entertainment catalogs; production agreements on behalf of music, television, and motion picture companies; and all entertainment issues relative to the Internet (Variety Staff 2013).

On March 5, 2012, I interviewed Jay L. Cooper about the new business of music and the implications for the movie industry. The following is a transcript of the video interview (Lee 2012):

Newton Lee How are the Internet and technological advances transforming the music business?

Jay L. Cooper Music has two ways to reach people: In person, when you're sitting in front of them listening to the music, or through all kinds of technology—it was radio, it was 78 records, it was originally piano rolls, then it became 78s, then it became 45s, and then it became 33s, and then it became CDs, and then it became DVDs, and now via the Internet, via satellite, and via cable. So music reaches the public through all these various sources. It's now reaching wirelessly, not by radio but by other means, as I said satellite. So technology has a major influence on how music is distributed, how music is sold, and you can't do music today without appreciating how you're going to deliver that music to the public.

Lee So basically, distribution has changed so much because of technology.

Cooper Distribution used to be that you would make a record, and if you weren't listening to it on the radio, and you wanted to buy a copy of it, you went to a store and you bought a copy of it. Today you don't have to go to a store to buy a copy, you can get it on the Internet, or you can get it by subscription service. There are various ways you can get music today. And of course, the idea of music is to reach the public, and the public wants access to the music, they want to be able to access the music that they like, whether they're at home or they're traveling, in a hotel room, on vacation, or in their office. They want to be able to get their music some way or somehow to listen to it.

Lee So the Internet also changes the marketing of music.

Cooper Marketing has always been difficult, but marketing was traditionally done in several ways: When a band or a singer went on tour—that was a form of marketing. When a sing or musician was interviewed on radio—that was a form of marketing. When they did a television show—that was a form of marketing. The concerts were a form of marketing. Then they would take ads out in newspapers or magazines, and that was another form of marketing. Now, marketing is using Facebook, using MySpace, and using YouTube in addition, and this is the new form of marketing. Still, people go out and perform, they go to the malls and they go to clubs and all that, but the new way of marketing reaching millions of people is through means like YouTube, through the Internet, through a webcast, so everything has changed in how you reach the public. Still, to this day though, television is still somewhat important. When somebody goes on *Saturday Night Live* or they go on the David Letterman show, they're reaching a lot of people, they're reaching millions of people. Or if they're on *The Voice* or *American Idol*, you're reaching people, and there are all these methods of marketing your talent and your product.

Lee So with faster internet and higher bandwidth, will the movie industry be facing the same fate as the music industry?

Cooper They are facing the same fate, if you're talking about the fate being piracy. Piracy is dramatically hurting our business in the following sense. More people are listening to music than ever before, and fewer people are buying music than ever before. What has happened is: It's all available on pirate sites, and people are just taking these and not realizing that a creator, musician, or songwriter's work product is music. That's what they do, that's what they provide to the public, and that's what they live off—the results of their work. But their incomes are being taken away.

The record industry since the year 2000 is down by 50 %. There were six major record labels and probably at least ten major independents. Now we are down to three major labels and maybe three to four major independents. There is less ability for an artist to find a way into the marketplace, because people are taking it rather than buying it. I don't know if the public always realizes that it's stealing, but it is. It's no different from walking into a record store, taking a CD and walking out with it. That's the same thing. That's what the musicians and singers and songwriters earn their living from: the sale of that product. The public doesn't realize that, and so, the record industry has suffered dramatically, and the public says, "Well, those rich record companies, and those rich artists..." We're not talking about the record companies here, we're talking about the thousands of artists who are not rich, that have earned their living and struggled for years to become successful. We're also talking about the money that the record companies have made off these successful artists, which they use to develop new artists. Out of what they put in ten different artists, maybe one would break through, because nobody knows who ultimately is going to be successful. You take a guess, but you never know who exactly is going to succeed, and the profits you get from successful artists is used to develop new artists, and what's happening now is that we have fewer and fewer artists being signed, so fewer opportunities for the public to discover new artists.

Well, now going to the motion picture industry: It's very expensive to do motion pictures. They can be $ 25, 100, or 200 million. $ 250 million is the cost of the new picture coming out now, called John Carter. Well, if somebody could steal those films, why would anybody invest $ 250 or 100 million or any dollars for that matter? What is happening is, for a long time, the profits that the motion picture companies were making, were off the DVD sales, because it's so expensive to produce a film. And once you produce a film, you have to go out and spend what they call "prints and advertising," and that costs millions and millions of dollars. Well, if your product is going to be stolen off of the Internet, you can't spend that money, and that means fewer films are being made. DVD sales are down over 50 %, and that's taken the profits right out of the movie business, so now they're facing the exact same problem. Sometimes these movies are out on the streets or on the Internet before they're even released in the theaters, and so they're suffering a terrible problem too.

Why the public doesn't entirely understand that this is wrong? It's the monies that one pays for the admission into a theater or from the purchase of a DVD or from advertising dollars on television that goes to finance the making of the movie. A movie doesn't involve just a star or two stars, it involves carpenters, electricians, lighting people, painters, and hundreds of people that it takes to make a film. All those people are losing their work because it's not just one or two stars, there are all the other actors and all the other background people that are needed to build sets or design costumes, to paint the sets, and to do all the things that are necessary to give you a good film. That takes hundreds and hundreds of people, and they're losing their jobs. So when we lose that expertise, we lose the filmmaking ability altogether.

Acknowledgements I would like to acknowledge my colleague Alan Kay (president of Viewpoints Research Institute) for introducing me to Quincy Jones; my intern Joey Lee (Ngee Ann Polytechnic) for his assistance in transcribing some of the video interviews presented in this chapter; and National University of Singapore and Media Development Authority for their wonderful internship program.

References

Consumer Electronics Association (2013) CE Hall of Fame Inductees. Consumer Electronics Association.   http://www.ce.org/Events-and-Awards/Awards/CE-Hall-of-Fame/Inductees.aspx. Accessed 21 Sept 2013

Ganim S, Stapleton A (2013) Dying teen covers Katy Perry's 'Roar'. CNN. http://www.cnn. com/2013/11/02/us/dying-teen-olivia-wise-roar/index.html. Accessed 4 Nov 2013

Lee N (2004a) Interviews with Quincy Jones. ACM Computers in Entertainment. http://dl.acm. org/citation.cfm?doid=973801.973815. Accessed Jan 2004

Lee N (2004b) Interviews with Karlheinz Brandenburg. ACM Computers in Entertainment. http:// dl.acm.org/citation.cfm?doid=1027154.1027169. Accessed July 2004

Lee N (2009) An Interview with Tom Silverman. ACM Computers in Entertainment. http://cie. acm.org/articles/interview-tom-silverman/. Accessed June 2009

Lee N (2012) How the internet and technological advances are transforming the music business: an interview with Jay Cooper. ACM Computers in Entertainment. http://cie.acm.org/articles/how-internet-and-technological-advances-are-transforming-music-business/. Accessed 5 March 2012

Variety Staff (2013) Transactional lawyers negotiate the deals that make showbiz grow. Variety. http://variety.com/2013/biz/features/transactional-lawyers-1200334425/. Accessed 16 April 2013

Wise O (2013) Olivia Wise—Roar (Katy Perry cover). YouTube. http://www.youtube.com/ watch?v=m_An8xNwupo. Accessed 14 Oct 2013

Chapter 2
Getting on the Billboard Charts: Music Production as Agile Software Development

Newton Lee

Computers and music are converging in a new era of digital Renaissance as more and more musicians such as will.i.am are learning how to code while an increasing number of software programmers are learning how to play music.

2.1 Music Appreciation and Songwriting

I dabbled with music composition before I learned computer programming. When I was in high school, one of my best friends, Kai Ton Chau, and I would go to the Hong Kong Arts Centre on the weekends and listened to hours of classical music. My appreciation of music grew from passive listening to active songwriting. For the high school yearbook, Chau and I decided to write a song together. Inspired by Rodgers and Hammerstein, he composed the music and I wrote the lyrics. The resulting sheet music was published in the yearbook.

Although I majored in electrical engineering and computer science during college years, my songwriting hobby did not dwindle over time. On the contrary, as soon as I landed my first full-time job at AT&T Bell Laboratories, I bought a professional Roland synthesizer and hooked it up to a Macintosh SE computer loaded with all the best music composition software at the time. I would write melodies and my vocal teacher Josephine Clayton would arrange the music.

As the founding president of Bell Labs' *Star Trek* in the Twentieth Century Club, I produced the first-ever "Intergalactic Music Festival" to showcase international songs and cultural dances performed by fellow AT&T employees. Indeed, music abounds in the *Star Trek* universe: Leonard Nimoy wrote and performed the song "Maiden Wine" in the original *Star Trek* episode "Plato's Stepchildren," and he

N. Lee (✉)
Newton Lee Laboratories, LLC, Tujunga, CA, USA
e-mail: newton@newtonlee.com

School of Media, Culture & Design, Woodbury University, Burbank, CA, USA
e-mail: newton.lee@woodbury.edu

N. Lee (ed.), *Digital Da Vinci,* DOI 10.1007/978-1-4939-0536-2_2,
© Springer Science+Business Media New York 2014

Fig. 2.1 The Disney Online "Love Crew": Mark Andrade, Bob Wyar, Michael Borys, Michael Bruza, Carrie Pittman, Cathy Georges, Frank Avelar, Gwen Girty, Eric Huff, Katie Main, Keir Serrie, Kenneth Ng, Mina Oh, Newton Lee, Pamela Bonnell, Robin Levey, and Elizabeth Swingle

played various musical instruments in "Charlie X," "The Way to Eden," and "Requiem For Methuselah" (Maiden Wine 2013).

Nimoy portrayed science officer Spock on board the *U.S.S. Enterprise*, who took pride in his emotionally detached, logical perspective. Yet, Spock displayed more musical talent than his crewmates. In reality, I have personally known many talented software engineers who are themselves gifted musicians: Henry Flurry, Harold Cicada Brokaw, Mark Tuomenoksa, Stephanie Wukovitz, and Jesse Gilbert, just to name a few.

Henry Flurry and Harold Cicada Brokaw are both award-winning music composers and software developers who worked with me on creating some of the bestselling Disney CD-ROM titles including *The Lion King Animated Storybook, Winnie the Pooh and the Honey Tree, Pocahontas, The Hunchback of Notre Dame,* and *101 Dalmatians* (Flurry 2010). Flurry, Brokaw, and I shared the 1995 Michigan Leading Edge Technology Award for creating Callisto—an innovative graphical authoring tool for multimedia CD-ROM software development.

Mark Tuomenoksa supervised me at AT&T Bell Laboratories on reverse engineering the Mac OS System 4.1 on the Apple Macintosh SE in the late 80's. Today, Tuomenoksa "spends his days developing everything from embedded algorithms for implantable medical devices to web platforms for autism research to IOS and Android applications that integrate biometric, environmental and observational monitoring. At night, Mark plays music, performing with bands at clubs and festivals, composing soundtracks and jingles, and playing sessions" (Tuomenoksa 2014).

Stephanie Wukovitz worked for me at Disney Online where she developed Java software code as well as composed background music for over 40 popular online games that were featured on Disney.com and the now-defunct DisneyBlast.com. We received many positive feedbacks from online players about the games as well as the music (Lee and Madej 2012). To celebrate the first anniversary of DisneyBlast. com in 1998, the Disney Online "Love Crew" wrote and recorded the song "Keep the Love Online," and I was chosen to be the lead rapper for the E. Huff and Puff Newton remix (Huff and Lee 2012) (see Fig. 2.1).

Jesse Gilbert is the chair of media technology at Woodbury University's School of Media, Culture, and Design. In 1994, Gilbert received a Watson Fellowship and spent a year studying music in Ghana, West Africa. His fascination with sound representation for tribal music and later heavy metal music led him to the development of SpectralGL—a visual instrument that employs an interactive software system to generate real-time 3D animation in response to live or recorded sound. Employing a number of computational analytical tools, including Fourier analysis and oscilloscope-style waveform deformation, SpectralGL reveals the deep structure of sound in a visual language that is both intuitively and aesthetically linked to our emotional experience of music. SpectralGL's interface gives the performer the means to generate highly dynamic 3D scenes that place the observer in a visual relationship that both enhances and reflects on the process of listening. In addition, SpectralGL's ability to overlay digital media onto its 3D surface expands upon traditional notions of presentation of the moving image, and of the relationship between sound and image (Gilbert 2013).

2.2   Relationship between Music and Computer Programming

What is the relationship between music and computer programming? Rob Birdwell explains in his blog dated November 2003, the same month of my Quincy Jones interview (Birdwell 2003):

- Creating music and software are simultaneously collaborative and individualistic undertakings.
- Musicians (regardless of era) are generally technically engaged - instruments themselves (the hardware) often interface with other devices (amps, mixers, mutes) to achieve different sounds. Composers often deal with an array of technologies to get their music written, performed and/or produced.
- Music is an abstract medium—the printed note requires interpretation and execution. Like the written line of code, there is often much more than meets the eye.
- Music is a form of self-expression. Many programmers (often to the dismay of corporate managers) try to express themselves through code.
- One famous music educator, Dick Grove, once said that composers/musicians often like to solve puzzles. (Dick Grove was very computer savvy - although I'm not sure he wrote code, I wouldn't doubt his ability to do so.)
- There is an infinite variety in music, musical ideas, and styles. Programmers are faced with a vast array of tools, concepts and languages for expressing and translating these ideas into something that achieves yet another result.

"Instrumentalists in particular (guitar players for example) make great programmers," Carl Franklin comments on Birdwell's blog. "It's not just about math and music being similar, or the fundamentals vs. the art. Instrumentalists have to zoom in to work with very repetitive technical details, and so become very focused—like a guitar player practicing a piece of music at a slow speed. But, the best programmers are able to then zoom out and see the big picture, and where their coding fits into the whole project, much like an artist has to step back from a painting and see the whole of it, or an instrumentalists has to produce something that communicates a complete work, not just the scales and technical aspects of it."

Nearly 10 years later in 2013, Birdwell wrote an update on his website (Birdwell 2013): "Both music and programming involve creating something from a very abstract concept or inspiration. Where programming and technology can be wondrous and grand (involving everything from a simple web site to collecting data from distant galaxies) so too can music. Both involve a 'commerce' aspect (i.e., commercial music, commercial software)—and yet both can be approached purely artistically. Both can be deeply personal and expressive."

Liz Ryan, CEO and founder of Human Workplace, opines that "Musical kids are smarter than most. They could major in anything—yet they choose music. If the kid decides to switch to biochemistry in mid-stream, I promise you, the kid will not fail. Music kids outscore all other majors in grad-school entrance exams. Why not let a kid with options pursue his musical dreams as far as that journey will take him? There's no downside" (Ryan 2013).

Given the intricate relationship between music and computer science, musical education can be useful for computer programmers, and software development is vice verse beneficial to musicians. This book presents the work of many researchers who wear at least two hats: a computer scientist and a musician. It shows how computer science has enabled practitioners to accomplish more than ever in the field of music.

A strong supporter of science, technology, engineering, and mathematics programs in schools, The Black Eyed Peas founding member will.i.am announced in July 2013 his future plan to attend the Massachusetts Institute of Technology to study computer science (McMohan 2013).

2.3   Computers and Music in the Digital Age

In 1961, an IBM 704 became the first computer to sing (Bell Laboratories 2008). According to U.S. Library of Congress, "This recording ["*Daisy Bell (Bicycle Built for Two)*"], made at Bell Laboratories on an IBM 704 mainframe computer, is the earliest known recording of a computer-synthesized voice singing a song. The recording was created by John L. Kelly, Jr., and featured musical accompaniment written by Max Mathews. Arthur C. Clarke, who witnessed a demonstration of the piece while visiting friend and Bell Laboratories employee John Pierce, was so impressed that he incorporated it in the novel and film script for '*2001: A Space Odyssey.*' When Clarke's fictional HAL 9000 computer is being involuntarily disconnected near the end of the story, it sings "Daisy Bell" as it devolves" (Library of Congress 2010).

The previous chapter "A Tale of Four Moguls: Interviews with Quincy Jones, Karlheinz Brandenburg, Tom Silverman, and Jay L. Cooper" chronicled the advancement of music technology since the forties, from stereo headphones to the Fender bass to synthesizers to DVD to the Internet. Music changes every step of the way as technology advances.

I remember listening to the radio in the eighties. Unlike the weekly Casey Kasem's Top 40 countdown, the radio DJs sometimes did not announce the names of the songs that they were playing. Therefore I had to use a cassette tape recorder to record bits and pieces of the songs that I like, play them back to the salespersons at some record stores who could help me identify the songs, and then purchase the singles or the entire albums on Vinyl or CD.

Computer hardware and software for music have improved drastically since. Shazam, SoundHound, musiXmatch, and Google Ears (Sound Search) are some of the mobile apps that can name that tune for you (Kelapure 2013). On iTunes, Google Play, and Amazon mp3, we can buy a single song without purchasing the entire album, and instantly without driving to a record store.

New musical instruments are being conceived and developed that will spawn new ideas and musical genres. The Artiphon, for example, is a multi-instrument (guitar, mandolin, bass, violin, and lap steel) powered by an iPhone or iPod Touch. "I wanted to make something that people at all skill levels could play, a device as agnostic to musical style as the piano but as expressive as a violin," said Mike Butera, inventor of the Artiphon (McNicoll 2013).

As for songwriting, today's composers fully embrace the latest computer technology in music production. In the chapter "Delegating Creativity" in this book, Prof. Shlomo Dubnov and Greg Surges wrote that "composers have increasingly allowed portions of their musical decision-making to be controlled according to processes and algorithms." In fact, computers and music are converging in a new era of digital Renaissance as more and more musicians such as will.i.am are learning how to code while an increasing number of software programmers are learning how to play music.

This book, in particular, holds dear to my heart because I have experienced firsthand as an executive producer overseeing the complete process of music production: from songwriting to recording to promotion. In 2011, four of the songs that I executive produced were aired on primetime TV (FOX and Lifetime). In 2013, one of the songs was charted on the U.S. Billboard Top 15 and U.K. Music Week Top 10.

2.4   Music Production (Software Development) Life Cycle

The Beatles producer George Martin once said, "A record producer is responsible for the sound 'shape' of what comes out. In many ways, he's the designer—not in the sense of creating the actual work itself, but he stages the show and presents it to the world. It's his taste that makes it what it is—good or bad." Oftentimes the right chemistry between recording artists and music producers is how hit songs are made.

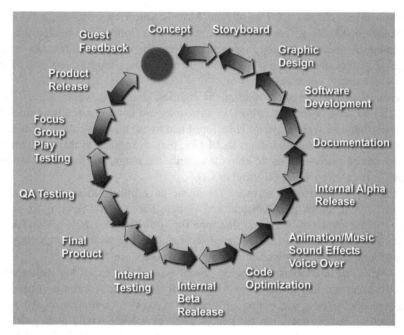

Fig. 2.2 Agile software development based on iterative and incremental life cycle

Indeed, M. Nyssim Lefford expounds producing and its effect on vocal recordings in the following chapter of this book.

While a music producer's job is to oversee the production process from start to finish, I as an executive producer put together a production team, manage the budget, lay out the timeline, and give the final approval of the musical products. The combined creativity of my team—consisted of music producers, songwriters, and performers—is evident in the music that we create. Since 2009, I have worked with many award-winning composers, producers, directors, and musicians who have their own different persona and creative styles. The challenge is to create a melting pot conducive to creativity in spite of their differences.

In every step of the music production, I manage the process according to agile software development based on iterative and incremental life cycle (see Fig. 2.2) while keeping in mind that making changes to a song or music video during its development cycle can be limited or cost-prohibitive.

2.4.1 Music Genre (Concept)

The first step is to decide on the genre of music, be it pop, dance, jazz, country, hip-hop/rap, R&B/soul, rock, alternative, or crossover music. The music genre determines who will be considered to join the creative team. In pop music, I have worked with Inessa Lee and Allyson Newman on "Write Me," "Remember Me," "Play with

Me," "Take Me to the Moon," and "Insane in F#." In pop-rap, I have collaborated with Ievgenii Bardachenko (JayB), Sjors Klaassen (DJ RMFH), and Princess X on "Summertime." In dance-pop, I have executively produced "You Turn Me On" with Mike Burns, Jonathan Reyes, and Princess X; "Gimme All (Ring My Bell)" and "Free" with L.C. Gonzalez, Tony Haris, Heidi Rojas, and Princess X; and "Dynamite" / "Динамит" with Liza Fox, Ana Sîrbu, and Radu Sîrbu.

2.4.2 Song Idea (Storyboard)

Is it an up-beat song or a sad song? Is it about love or a charitable cause? Such questions are raised in the brainstorming sessions. Inspiration is the most important ingredient at this stage. For example, British musician Jerry Dammers wrote the song "Free Nelson Mandela" (1984) that became the anti-apartheid anthem (Simpson 2013). Released on July 4, 2013, the new single "Free" was inspired by civil rights activist Martin Luther King, Jr., and the song contained samples licensed from the 1991 classic "Everybody's Free (To Feel Good)" (16).

2.4.3 Singer Image (Graphic Design)

What kind of image will the singer project with the song? A light and carefree image or a dark and sophisticated image? The performers and the producers have the biggest input into the singer image. The image will also influence the look and feel of the music video that will be created for the song. For instance, the Betty Boop image greatly influenced Helen Kane's signature song "I Wanna Be Loved by You" in 1928.

2.4.4 Songwriting: Melody, Lyrics, Arrangements (Software Development)

Having chosen the music genre, come up with a song idea, and visualized the singer image, it is time to start writing the song. Many singers are also songwriters who actively participate in the songwriting sessions. I have attended many creative sessions where producers, singers (including background vocalists), songwriters, and I have all contributed to the melody and lyrics of a new song. There are two common strategies in composing a new song: Start with a melody and add an arrangement, or start with an arrangement and come up with a melody.

2.4.5 Songwriting Documentary (Documentation)

During the entire creative process, we compare notes, record bits and pieces of melody, discusses other existing songs, take pictures, et al. The information can easily be turned into a documentary.

2.4.6 Initial Draft (Internal Alpha Release)

The initial draft of the song is completed at this stage. A temporary vocal track may be added to it. The creative team listens to the initial draft and suggests changes to the melody, lyrics, and instrumentals.

2.4.7 Tracking: Recording Session (Animation/Music, Sound Effects, Voice Over)

Once the creative team is satisfied with the first complete draft, we would record the singers and background vocalists at a professional studio such as Peermusic in Burbank, Paramount Recording Studios in Hollywood, and The Invisible Studios in West Hollywood. Melodies, harmonies, overdubs, background vocals, and live instrumentals (if necessary) are meticulously recorded. Akin to filmmaking, music producers act like directors, and singers perform as if they were actors. Recording live performances over and over, the team aims to perfect each "take", record the best sonic quality possible, and experiment with some variations on different takes.

2.4.8 Mixing (Code Optimization)

Now that we have all the recorded sounds, it is time to blend the arrangement and the multiple recorded elements together into a final recording that is most appealing to listeners. Mixing controls what is coming out of the end listener's speakers: Bass becomes floor shaking, kicks begin to hit you in the chest, and vocals soar with shimmering reverb and echoes. The music producer and sound engineer often use equalizers, compressors, faders, reverbs, echoes, ear candies, and auto-tune to add space, depth, and color to the sound. Cher's 1998 hit song "Believe" was the first major commercial success featuring the extensive use of auto-tune (Frere-Jones 2008). R&B singer T-Pain and many pop artists have further popularized auto-tune for both pitch corrections and special effects.

2.4.9   Final Draft (Internal Beta Release)

The final draft of the song contains all the elements of the vocals, instrumentals, and ear candies. Unless the creative team is completely satisfied with it, we may demand additional fine-tuning or even a second follow-up recording session if necessary.

2.4.10   Remixes (Internal Testing)

During this internal testing period, the multi-track stems including vocals and instrumentals are sent to additional producers and DJs who would remix the song into other music formats such as dubstep, progressive house, techno, and extended play for dance clubs. Remixes are a very common tool for dance music artists to gain a bigger audience worldwide. Multiple remixes playing simultaneously also helps push a song up the Billboard charts. I have collaborated with the Almighty, Mark Picchiotti, Ralphi Rosario, and the Hoxton Whores (Kevin Andrews and Gary Dedman) on the remixes for "Gimme All (Ring My Bell);" and Jason Donnelly (DJ Puzzle), Sven Erler, Daniel Gardner (DMG), Håkan Hannu, Nyxl, Joe Rare, Ilonka Rudolph, and James Sharman (Grixis) on the remixes for "You Turn Me On."

2.4.11   Audio Mastering (Final Product)

With a minimum of 3 to 6 dB of available headroom, the final song and remixes are sent to mastering engineers who would apply corrective equalization and dynamic enhancement in order to optimize the sound on all playback systems. To master an album consisting of multiple songs, mastering engineers would re-balance each song to create a cohesive sound for the entire album. It is important to have an acoustically-treated and sound-proofed environment for superior listening. Universal Mastering Studios, a division of Universal Music Group, is one of the popular places for audio mastering. Grammy award winning engineer Erick Labson, for instance, mastered Liza Fox's "Dynamite" / "Динамит" to give the song more oomph.

2.4.12   Variations of a Mastered Song (QA Testing)

There are times when the mastering engineers would give us a few variations of the mastered song to consider (for example, warmer versus brighter versions). The creative team would choose the best mastered version as the final product.

Latest	Archive

Commercial Pop 28/01/2013 to 03/02/2013 (Week 5)

● New Entry   ● Re Entry   ● Highest New Entry   ● Highest Climber   ★ Award

✉ Contact   🖨 Print   📥 Download

TW	LW	Wks	Artist/Title	Label	Mixes
1	5	2	**EXAMPLE** Perfect Replacement	MoS	(R3hab & Hard Rock Sofa/Datsik/Toyboy & Robins/Danny Howard Mixes)
2	6	4	**LAWSON** Learn To Love Again	Global Talent/Polydor	(Cutmore/Noise Freakz/Jumpsmokers Mixes)
3	20	2	**JUSTIN TIMBERLAKE FEAT. JAY-Z** Suit & Tie	RCA	
4	1	5	**AVICII V NICKY ROMERO** I Could Be The One	Positiva Virgin	(Original/Audrio/Didrik Mixes)
5	15	3	**JESSICA WRIGHT FEAT. MANN** Dominoes	AATW	(Rudedog/Chuckie Mixes)
6	11	3	**RITA ORA** Radioactive	Columbia/Roc Nation	(Lucien Fort/Baggi Begovic/Zed Bias/Waze & Odyssey Street Tracks/The Flexican Mixes)
7	21	2	**KIMBERLEY WALSH** One Day I'll Fly Away	Decca	(The Alias Mix)
8	13	4	**BETH SHERBURN** Ordinary World	White Label	(Andi Durrant & Steve More/Rob Searle/Wonky/Flash Trash Mixes)
9	14	6	**PRINCESS X** Gimme All (Ring My Bell)	White Label	(Ralphi Rosario/Mark Picchiotti/Almighty/Hoxton Whores Mixes)

Fig. 2.3 U.K. Music Week commercial pop chart (January 28–February 3, 2013)

2.4.13 Feedback from Professionals and DJs (Focus Group Play Testing)

At our discretion, we may send out the song and its remixes to our close friends, DJs, and promoters for their feedback. At this point, we may work on a music video for the synchronous release of the song. A music video director interprets the song

Fig. 2.4 U.S. Billboard dance/club play songs chart (April 13, 2013)

and creates some compelling visual elements for storytelling. I have worked with music video director Thomas Mignone on "Take Me to the Moon," Phil Lee on "Gimme All (Ring My Bell)," David Shawl on "You Turn Me On," Pia Klaar on the Second Life filming of "You Turn Me On (Machinima)," and Bakhodir Yuldashev on "Dynamite" / "Динамит."

2.4.14 Song Release (Product Release)

This is the big day when we finally release the new song, its remixes, and the music video to the public through iTunes, Spotify, YouTube, Facebook, MySpace, Amazon.com, and other distribution channels.

April 13 2013 billboard

Dance/Electronic

DANCE/ELECTRONIC SONGS™

2 WKS AGO	LAST WEEK	THIS WEEK	TITLE / PRODUCER (SONGWRITER)	Artist / IMPRINT/PROMOTION LABEL	CERT	PEAK POS	WKS ON CHART
1	1	1	**HARLEM SHAKE** BAAUER (H.RODRIGUES)	Baauer JEFFREES/MAD DECENT/WARNER BROS.		1	8
3	2	2	**FEEL THIS MOMENT**	Pitbull Feat. Christina Aguilera MR.305/POLO GROUNDS/RCA		2	12
2	3	3	**SCREAM & SHOUT** LAZY JAY (W.ADAMS,J.MARTENS,L.BAPTISTE)	will.i.am & Britney Spears INTERSCOPE	▲	1	12
5	4	4	**SWEET NOTHING** C.HARRIS (C.HARRIS,F.WELCH,A.HARPOON)	Calvin Harris Featuring Florence Welch DECONSTRUCTION/FLY EYE/ULTRA/ROC NATION/COLUMBIA		3	12
7	6	5	**I LOVE IT** PATRIK BERGER (P.BERGER,L.EKLOW)	Icona Pop Featuring Charli XCX RECORD COMPANY TEN/BIG BEAT/RRP		5	12
4	5	6	**DON'T YOU WORRY CHILD** AXWELL,S.INGROSSO,S.ANGELLO (AXWELL,S.INGROSSO,S.ANGELLO)	Swedish House Mafia Feat. John Martin ASTRALWERKS/CAPITOL	▲	2	12
6	7	7	**GANGNAM STYLE** PSY,U-HYUN (PSY,PARK,J.HYOO)	PSY SCHOOLBOY/REPUBLIC	▲	4	12
8	8	8	**ALIVE** RAIN MAN (J.YOUSAF,J.YOUSAF,K.TRINDL,N.LIM,J.UDELL)	Krewella KREWELLA/COLUMBIA		8	12
—	9	9	**#THATPOWER** D.LEROY,WILL.I.AM (W.ADAMS,D.J.LEROY,J.BIEBER)	will.i.am Featuring Justin Bieber INTERSCOPE		8	2
9	13	10	**LEVITATE** J.DAG'STAR (HADOUKEN,N.SMITH,N.HILL,C.HARVEY)	Hadouken! SURFACE NOISE		9	11
12	10	11	**CLARITY** ZEDD (A.ZASLAVSKI,MATTTHEW KOMA,P.ROBINSON,S.GRAY)	Zedd Featuring Foxes INTERSCOPE		8	12
11	11	12	**I COULD BE THE ONE**	Avicii vs Nicky Romero LE7ELS/CASABLANCA/REPUBLIC		10	10
10	12	13	**AS YOUR FRIEND** AFROJACK (N.VAN DE WALL)	Afrojack Featuring Chris Brown WALL/ISLAND/IDJMG		8	10
14	15	14	**BEAM ME UP (KILL-MODE)** A.BJORKLUND,S.FURBER (K.SHEEHAN,A.POURNOURI,A.BJORKLUND,S.FURBER)	Cazzette AT NIGHT		14	12
13	14	15	**FOREVER NOW**	Ne-Yo MOTOWN/IDJMG		12	11
20	17	16	**I NEED YOUR LOVE** C.HARRIS (C.HARRIS,E.GOULDING)	Calvin Harris Featuring Ellie Goulding CHERRYTREE/DECONSTRUCTION/FLY EYE/ULTRA/ROC NATION/INTERSCOPE/COLUMBIA		16	12
16	16	17	**SPECTRUM** ZEDD (A.ZASLAVSKI,MATTTHEW KOMA)	Zedd Featuring Matthew Koma INTERSCOPE		10	12
—	26	18	**HIGHER GROUND** R.RICHARD,L.F.PIERRE II (R.BIRCHARD,L.F.PIERRE II)	TNGHT WARP		18	2
19	18	19	**SHE WOLF (FALLING TO PIECES)** D.GUETTA (D.GUETTA,S.FURLER,C.BRAIDE,G.H.TUINFORT)	David Guetta Feat. Sia WHAT A MUSIC/VIRGIN		8	12
15	21	20	**GET UP (RATTLE)**	Bingo Players Feat. Far East Movement SPINNIN'/CASABLANCA/BUG		15	5
22	19	21	**CALL ME A SPACEMAN** HARDWELL (N.VAN DE CORPUT,M.CROWN)	Hardwell Featuring Mitch Crown CLOUD 9		19	9
18	20	22	**REST OF MY LIFE** D.GUETTA,G.TUINFORT (C.J.BRIDGES,U.RAYMOND IV,J.SALAAS, R.D.SKLAR,W.SCANDRICK,D.GUETTA,G.TUINFORT)	Ludacris Featuring Usher & David Guetta DTP/DEF JAM/IDJMG		6	12
23	22	23	**HOLD ME** D.AUDE (D.AUDE,X.OXID)	Ono Featuring Dave Aude MIND TRAIN/TWISTED		22	7
24	24	24	**RIGHT NOW**	Rihanna Featuring David Guetta SRP/DEF JAM/IDJMG		19	12
—	34	25	**GLOWING** SANDY VEE,DREAM_AB,A.BIRGISSON (A.BIRGISSON,L.HAYWOOD,D.JAMES,B.REXHA,S.WILHELM)	Nikki Williams ISLAND/IDJMG		25	2
25	25	26	**SEXY PEOPLE (THE FIAT SONG)** O.FAWOLO,J.RIBA (D.REDOVA (AZ PEREZ),J.RIBA,J.KESHAW, O.P.FO,O.REGAN,O.OLAJIDE,L.GOMEZ,J.GARCIA,J.AMANDEES)	Arianna Featuring Pitbull ULTRA		24	7
42	29	27	**BACK TO LOVE** J.REMULLO,PAULY P (J.DEGIYOO,D.JAY SEAN,J.COTTER,J.SGALKAY,R.DHANOA,N.HUNNIFEN,S.HUMMEDES)	DJ Pauly D Featuring Jay Sean G NOTE/US UNIT		7	8
28	27	28	**FOREVER** R.ROSARIO (R.ROSARIO,C.MORROW,N.P.PHILLIPS O.LAND)	Ralphi Rosario Featuring Frankie RAMARO		27	6
36	31	29	**ACID RAIN** STARGATE,BINGO PLAYERS (S.FURLER,M.S.ERIKSEN,T.E.HERMANSEN)	Alexis Jordan STARROC/ROC NATION/COLUMBIA		29	4
27	28	30	**LOUDER** D.STEIN,S.EVANS (D.STEIN,S.EVANS)	DJ Fresh Featuring Sian Evans COLUMBIA		21	11
35	32	31	**APOLLO** HARDWELL (N.VAN DE CORPUT,A.SHEPHERD)	Hardwell Featuring Amba Shepherd REVEALED/CLOUD 9		23	11
32	33	32	**GIMME ALL (RING MY BELL)** L.C.GONZALEZ (J.LEE,N.ROJAS,L.C.GONZALEZ,F.KNIGHT)	Princess X IONESA LEE		32	5
26	30	33	**OH MAMA HEY** C.COX,F.ANOBILE (C.COX,F.ANOBILE,C.WATERS)	Chris Cox + DJ Frankie Feat. Crystal Waters TOMMY BOY		17	9
30	35	34	**DRINKING FROM THE BOTTLE** C.HARRIS,D.REYNOLDS,A.WRIGHT (C.HARRIS,D.KASHAL,D.REYNOLDS,M.WRIGHT)	Calvin Harris Feat. Tinie Tempah DECONSTRUCTION/FLY EYE/ULTRA/ROC NATION/COLUMBIA		20	12
29	36	35	**DOWN THE ROAD** 2DEVI. (S.RICHARD,G.HULIN,A.FRADIN,P.FORESTIER,CLE VEXIER)	C2C ON AND ON/CASABLANCA/REPUBLIC		28	11
HOT SHOT DEBUT		36	**I LOVE IT** NOT LISTED (P.BERGER,C.AITCHISON,L.EKLOW)	Melissa Adams TAUCHER		36	1
—	40	37	**X YOU** AVICII,A.POURNOURI (T.BERGLING,A.POURNOURI)	Avicii LE7ELS/CASABLANCA/REPUBLIC		37	2
44	45	38	**ONE MINUTE** RAIN MAN (J.YOUSAF,J.YOUSAF,K.TRINDL)	Krewella KREWELLA/COLUMBIA		38	5
NEW		39	**CHASING SUMMERS** TIESTO,S.V.ERNEST,SHOWTEK (T.V.ERMEEST,S.JANSSEN,W.JANSSEN)	Tiesto MUSICAL FREEDOM		39	1
37	38	40	**JUST ONE LAST TIME** D.GUETTA,G.TUINFORT (D.GUETTA,G.ASGIER,O.LILA,DGEN,A.GUETTA,G.H.TUINFORT)	David Guetta Featuring Taped Rai WHAT A MUSIC/ASTRALWERKS/CAPITOL		29	12
NEW		41	**I LOVE IT** NOT LISTED (P.BERGER,C.AITCHISON,L.EKLOW)	Hit Mix ICOVER		41	1
38	43	42	**YEARS** ALESSO,MATTHEW KOMA (A.LINDBLAD,MATTTHEW KOMA,S.WATTERS)	Alesso Featuring Matthew Koma REFUNE/CASABLANCA/REPUBLIC		31	7
34	39	43	**CRYSTALLIZE** MARKO G. (L.STIRLING,M.GLOGOLJA)	Lindsey Stirling BRIDGETONE		34	10
RE-ENTRY		44	**PLAY HARD** D.GUETTA (D.GUETTA,G.H.TUINFORT,C.RIESTERER,A.THIAM,S.C.SMITH,S.MOLIJN,K.N.SEKERG)	David Guetta Featuring Ne-Yo & Akon WHAT A MUSIC/VIRGIN/EMI		28	7
39	44	45	**BRING OUT THE BOTTLES** REDFOO (S.K.GORDY,B.GARCIA,A.SMITH)	RedFoo FOO & BLU/CHERRYTREE/INTERSCOPE		36	12

Fig. 2.5 U.S. Billboard dance/electronic songs sales chart (April 13, 2013)

Fig. 2.6   A behind-the-scenes photo for the Princess X music video "Gimme All (Ring My Bell)" on May 7, 2012

2.4.15   Feedback from Broadcast Industry and the Public (Guest Feedback)

On February 9, 2011, three original songs—"Write Me", "Remember Me," and "Play with Me"—were prominently featured on the FOX TV reality show *American Idol* (season 10) (IMDb 2011). On November 5, 2011, "Insane in F#" was aired on in the Lifetime TV original movie *Cheyenne* (a part of *Five [for the Cure]*) directed by Penelope Spheeris (Lifetime 2013).

The U.K. Music Week and U.S. Billboard magazines are among the most coveted music industry publications. "Gimme All (Ring My Bell)" peaked at #9 on the U.K. Music Week Commercial Pop Chart in January 2013 and stayed for a total of seven weeks on the chart (see Fig. 2.3). The song also peaked at #12 on the U.S. Billboard Dance/Club Play Songs Chart in April 2013 and stayed for a total of nine weeks on the chart (see Fig. 2.4), without resorting to questionable tactics employed by some big labels to push a song up a DJ chart (Karp 2013). The public response was equally well-reflected in the Billboard Dance/Electronic Sales Chart as the song was ranked #32 in terms of music sales in April 2013, surpassing some of the new songs by other big-name producers and artists including Avicii, David Guetta, and Tiesto (see Fig. 2.5).

The music video for Princess X's "Gimme All (Ring My Bell)" has garnered over 1 million views on YouTube in 2013 (see Fig. 2.6 for a behind-the-scenes photo). In a follow-up to the success, a music video was filmed for Liza Fox's "Dynamite" / "Динамит" in March 2014. The singer herself performed the entire

Fig. 2.7 On location shooting of the Liza Fox music video "Dynamite" / "Динамит" on March 4, 2014

stunt inside and outside a moving vehicle that was captured on a 4K RED camera mounted on a Russian Arm (see Fig. 2.7).

Acknowledgements Often waking up in the morning with a random song playing in my head, I am thankful for the creative minds that make good music in many genres and languages. Moreover, I would like to acknowledge Kelly Farrell, Steve Gilmore, Craig Jones, Brad LeBeau, DJ Dan de Leon, DJ Dave Matthias, DJ Stephanie Swanson, DJ Michael Toast, Sàndor Von Mallasz, Matt Waterhouse, DJ Will, and many other fans and supporters.

References

Bell Laboratories (2008) First computer to sing—Daisy Bell. YouTube. [Online]. http://www.you-tube.com/watch?v=41U78QP8nBk. Accessed 9 Dec. 2008
Birdwell R (2003) The relationship between programming and music. [Online]. http://weblogs. asp.net/rbirdwell/archive/2003/11/14/37643.aspx. Accessed 14 Nov. 2003
Birdwell R (2013) The relationship between programming and music—Part 2. [Online]. http:// birdwellmusic.com/articles/the-relationship-between-programming-and-music-part-2/. Accessed 4 Aug. 2013
Flurry H Henry Flurry (2010) [Online] http://www.henryflurry.com/
Frere-Jones S (2008) THE GERBIL'S REVENGE: Auto-Tune corrects a singer's pitch. It also distorts—a grand tradition in pop. [Online] The New Yorker. http://www.newyorker.com/arts/critics/musical/2008/06/09/080609crmu_music_frerejones?currentPage=all. Accessed 9 June 2008

Gilbert J (2013) SpectralGL. [Online] Jesse Gilbert. http://jessegilbert.net/code/spectralgl/. Accessed 29 Aug. 2013

Huff E, Lee N (2012) Disney stories: keep the love online. [Online] YouTube. https://www.youtube.com/watch?v=pLRnQDUtlU0. Accessed 18 Aug. 2012

I.L.M (2013) "I Have A Dream" (Martin Luther King Jr.) #DreamDay. *YouTube*. [Online] http://www.youtube.com/watch?v=NR_81O7atxI

IMDb (2011) American Idol: Season 10, Episode 7. [Online] IMDb. http://www.imdb.com/title/tt1830927/. Accessed 9 Feb. 2011

Karp H (2013) Funky Remix? How to reach No. 1 on billboard's DJ chart. [Online] The Wall Street Journal. http://online.wsj.com/article/SB10001424127887323689204578573981625735410.html. Accessed 8 July 2013

Kelapure A (2013) Best app for identifying music: shazam, soundhound, or MusixMatch? [Online] evolver.fm. http://evolver.fm/2013/04/12/best-app-for-identifying-music-shazam-soundhound-or-musixmatch/. Accessed 12 April 2013

Lee N, Madej K. (2012) Disney stories: getting to digital. [Online] Springer. http://www.amazon.com/Disney-Stories-Getting-Newton-Lee/dp/1461421004. Accessed 27 April 2012

Library of Congress (2010) The sounds of fighting men, howlin' wolf and comedy icon among 25 named to the national recording registry. U.S. Library of Congress. [Online]. http://www.loc.gov/today/pr/2010/10-116.html. Accessed 23 June 2010

Lifetime (2013) FIVE. [Online] Lifetime. http://www.mylifetime.com/movies/five. Accessed 5 Oct. 2013

McMahon, Sean. (2013) will.i.am acts like a sponge to promote STEM education. [Online] SmartBlog on Leadership, http://smartblogs.com/leadership/2013/07/09/will-i-am-acts-like-a-sponge-to-promote-stem-education/. Accessed 9 July 2013.

McNicoll A (2013) Artiphon instrument 1: guitar, keyboard, drums and bass, in one place. [Online] CNN. http://www.cnn.com/2013/06/27/tech/innovation/artiphon-instrument-1-iphone/index.html. Accessed 27 June 2013

Maiden Wine. (2013) The Musical Touch of Leonard Nimoy. [Online] Maiden Wine. http://www.maidenwine.com/catalog_05_film_stage.html. Accessed 1 Oct. 2013

Ryan L (2013) Let the kid study music, already! [Online] LinkedIn. http://www.linkedin.com/today/post/article/20130728144536-52594-let-the-kid-study-music-already. Accessed 28 July 2013

Simpson D (2013) Jerry Dammers: how I made Free Nelson Mandela. The Guardian. [Online]. http://www.theguardian.com/music/2013/dec/09/jerry-dammers-free-nelson-mandela. Accessed 9 Dec. 2013

Tuomenoksa M (2014) Will Power Blues. [Online] www.willpowerblues.com. Accessed 19 Feb 2014

Chapter 3
Producing and Its Effect on Vocal Recordings

M. Nyssim Lefford

This chapter investigates one of the more enigmatic aspects of record production: producing. In particular, it considers how producers influence the sonic attributes of vocal recordings and how technology intersects with the act of producing. The producer's actions, though they remain poorly understood, have long been deemed a necessary part of music recording. Producers advise and sometimes outright direct the creation of "the sound" of a recording, meaning they shape not one particular performance or aspect it but each aspect so all the different sounds fit together. Producers work on individual parts, but with the whole always in view. This attention leads ideally to a coherent listening experience for the audience. Some parts are exceedingly ephemeral, such as the attributes of the recorded voice. Yet the voice, with its exceptional malleability, provides essential glue for piecing together coherent wholes.

The nature and function of these vocal attributes and even more generally the attributes that constitute "the sound" of a recording remain topics for debate and research. The importance of the producer's contribution to their creation, however, is uncontested, and extensively documented in music history books and the sales charts of the recording industry's trade press. Producers were present in the earliest recording sessions (e.g. Fred Gaisburg and Walter Legge), and there is no hint of them disappearing soon. To the contrary, the role of the producer in recording is evolving and expanding. According to Frith, rock criticism reveals the indicators of change. "If producers were always implicitly involved in the rock story, their role is now being made explicit." (Frith 2012, p. 221) Increasingly, producers are credited as the central artistic force in productions, and increasingly, musicians self identify as producers and/or self-produce. For these reasons alone, producing deserves greater attention. In addition, technology is pressing the issue. New technologies enable but also encroach on existing production methods and techniques.

The producer's expertise is not only ill defined, but it also varies greatly among practitioners; some producers being more musically literate, others more technical, some very business minded, others almost shamanistic in their approach to leading

M. N. Lefford (✉)
Luleå University of Technology, Luleå, Sweden
e-mail: nyssim.lefford@ltu.se

N. Lee (ed.), *Digital Da Vinci*, DOI 10.1007/978-1-4939-0536-2_3,

sessions. Whatever the mysterious purpose producing serves and whatever methods it utilizes, it achieves (strives to achieve) captivating recordings. Given the diversity of approaches and the tremendous variety among recordings that captivate, the mechanisms by which this affect is achieved are not simple to detect. What consistencies do producers and the varied modes of producing share? What sonic qualities in performances and recordings attract the producers' attention, regardless of how these sounds are to be individually acted upon? This survey deconstructs the process of producing to begin to uncover how produced recordings catch attention, first the producers' and then other listeners, and how producing exudes influence over audiences.

The voice is singled out as an essential part of the whole, a primary and primordial instrument that calls to listeners on many levels and whose features other instruments often imitate. It will be discussed why producing exerts influence over vocal performances in particular, and why some vocal affectations may be more compelling to listeners, and thus of more interest to producers. Finally, approaches to the analysis of the producing "effect" are discussed to suggest inroads for both producers and technologists interested in acting on this effect. Future creative, technical and scientific developments in the art of production will emerge from a deeper understanding of the producing process and an appreciation of its numerous facets.

The producers' actions and decisions are guided by multifarious aesthetic, technical, logistical and perceptual considerations. Some decisions are motivated by philosophy, others by something resembling a designed method. To offer a purely technical view that minimizes the artistic one or to discuss the producer's relationship with musical creators but not with the listening audience would provide a very skewed, unrealistic perspective of what producing does. Producing apportions due consideration appropriately, for each production, so that the concerns of each vantage inform how a meaningful communication is created. Correspondingly, media and (computer) music studies, cognitive science and biology all provide frameworks through which producing may be understood as well as methodologies for its study. Appropriately, some of these disciplines deal with philosophical matters, while others with scientific evidence.

The chapter has four main sections. Each integrates varied sources and findings from across fields to highlight areas of commonality. It starts with an overview of pertinent processes and models in music and music recording. Then the material digs deep into producing itself, what is production's effect and what does this effect do to vocal performances in particular. Next, these findings are compared to other modes of (non-linguistic, non-verbal) communication, and viewed from the vantages of perception and socio-biology. This assessment helps to explain why the effect of producing is so intensely affecting to listeners, and also, serves to identify detectable, measurable patterns in the producing process. The chapter concludes by enumerating a set of approaches for deconstructing producing, productions and potentially the work of specific producers, and the applications of such findings to new tools and methods for production.

3.1   Process and Procedures

Many of the roles producers play in sessions have remained consistent and constant even though, from a purely technical point of view, the recording process itself has undergone many dramatic changes. The most influential of these innovations have come from computational advances, which have impacted every part of the recording chain from the artist emoting to the distribution of the final product. Each jump in processing power has sparked shifts in recording practices; and not least important among them is an unprecedented reliance on algorithms to enhance and optimize recorded audio signals. Now prevalent digital signal processors and plug-ins "normalize", "denoise", add "warmth", add "punch", adjust pitch, quantize, distort, improve or fix in various ways recorded signals. Extinct analog devices are resurrected virtually, making it simple to imitate recording techniques from bygone eras even under the most modern production conditions. Still other tools offer the possibility to "humanize" sequenced/programmed performances. In nearly every recording studio, bundled together with digital audio workstations, there exist enumerable types of digital emulations and emulators that have taken over, or may soon take over, processes integrally connected with how recorded music is produced.

Since both the technical and human systems involved in music production are complex, recording has traditionally been a collaborative endeavor in which responsibilities for the requisite expertise have been distributed among participants. The division of labor among artists, engineers and producers allowed contributors to focus on areas best suited to their expertise. Furthermore, as recording technology has been until fairly recently expensive and exclusive, its functionality and application were understood by a few specialists. This, however, has changed. Continually, technology becomes cheaper and more accessible. New designs emphasize ease of use for novices and semi-professionals. And, this very same technology is becoming smarter, making it possible to delegate some of the responsibility for requisite expertise to machines. In the modern digital recording studio, even the most mundane tools—or especially the most mundane tools—used for the most common of tasks *assist* the user. The greater the algorithmic sophistication of these technologies, the greater the autonomy they and/or their software developers enjoy during the production process. There is cause to anthropomorphize, since sonic creators increasingly rely on the procedural expertise represented by these devices instead of human collaborators. In this way, computation overtakes traditional practices in which human recordist make decisions about how technical procedures apply and are executed.

This progress need not be approached with foreboding, but it does require honesty. Computation is a tremendously powerful tool for artistic expression. Software is programmed to execute procedures compliantly (whether with a humanistic touch or not). The danger lies in a lack of understanding the procedures these tools execute. For example, the sound engineer or record producer who does not know anything about how a digital signal is made to sound like it is being replayed off of analog tape will be unable to make fine discriminations about the effect, about the type of tape being emulated, its imaginary width or ips (inches per second), etc. If a recordist is unable to discern beyond sounding cool/uncool, then the software emulation and software designer, as ghost producers, make these finer discriminations in the design of their algorithms (before releasing *their product*), and/or demarcate through the plug-in's control interface what discriminations may be made in its use. High-level users/creators in any digital media confront difficulties understanding, controlling and operating underlying systems behavior. Mateas cautions that in the absence of understanding what the code in digital media represents and does, without "procedural literacy", the meaning of using specific procedures in specific contexts is lost to those who would study digital media artifacts. (Mateas 2005) As recording technology grows more and more sophisticated, the distinctions between the content, an app that delivers it and the tools we use to create these things blurs and artistic intention reveals itself less and less directly; which should be of concern to the producer, who is concerned with the formulation artistic ideas.

The Intention to Imitate All creative tools (computational or otherwise) have characteristics that cause attributes to appear in what they are employed to render. The skill to use any tool presupposes an understanding of how its usage leads to specific attributes. Digital technology and algorithms are in no way the enemy of control; but in recording, processes that were once under more direct control of humans, that were more flexibly parameterized and adjusted to suit the particular needs of each scenario, are now more often left to algorithms... with presets. In some cases, tools fully automate the execution of engineering and production choices by globally adjusting (as prescribed by mathematics not perception or aesthetics) an entire track or mix (e.g., normalize levels, quantize performances). What is creative decision making in this scenario? Because the algorithms make their mark with calculated indifference, the more opaque their operations, the less a user is able to exercise exacting technical control and creative intention over what is rendered. In the recording studio, for producers or sounds engineers to *mean* for something to sound a particular way, to formulate with intention, they must have command over what the effects and affects do to the sound and what that sound does to listeners. Algorithms challenge deep-seated notions of expertise, the division of labor among experts and the power to delegate responsibilities for production decisions.

It must be reiterated, for the thorough creator, the ability to represent recording procedures algorithmically presents far, far more opportunities than liabilities. There is the potential for overall improvements to sound quality technically speaking. The functionality in plug-ins, according to some criteria, has been optimized. For managing complexity, the ability to automate complex procedures is an asset that potentially encourages greater attention to detail and more nuanced productions. Computation makes possible creative choices impossible without computation and/or by freeing up time and resources for aesthetic exploration. To unlock the desirable possibilities of computation, the connections between human processes and algorithmic procedures need to be better understood. Fortuitously, in this the technology itself provides valuable tools for understanding. Because we find algorithms already crossing into territory that previously has been under the watch of producers, the domain of giving a sound a "sound", the technology can provide a useful lens for magnifying the sorts of decisions producers made historically.

Historically speaking, although computation has not been the framework, it would be inaccurate to say that production has ignored the value of codifying procedures to emulate particular sounds. Producers since the earliest days of recording, and without algorithmic aid, have intentionally mimicked sounds and patterns from other recordings. Old sounds, or sounds similar to old sounds, were re-created for new contexts. Processes of a technical, creative and musical nature and sonic artifacts all are still routinely imitated and therefore are being analyzed, parameterized and procedural-ized. In a way, the producer's expertise overlaps with cognitive science and generative systems and emulator design, although the producer's methods tend to be qualitative not quantitative and informal though frequently no less systematic. Evidence of this practice is to be found everywhere. Producers, engineers and musicians commonly reference recordings in discourse, during production and in collaboration, both the sounds of them and the performances in them. Completed recordings provide touchstones and inspiration for new productions. It is a widely recognized cliché and almost vulgar to mention that many pop producers endeavor to follow a hit recording with something that sounds similar but fresh. The commercial mindset begets a formulaic approach. Mainstream commercial music production seeks to originate even while remaining firmly rooted in the expected, predictable features of a style that has proven popular. Whether the producer's goal is to avert risk or innovate, the choice of which formula, of which features to reference or mimic, is crucial. Overt imitation appears unoriginal, manufactured, calculated and disingenuous. Something with no discernable points of reference leaves listeners adrift, and is unlikely to attract much of an audience. The knowledge required to choose features, although poorly documented, perpetuates as shared intuition if not as outright praxis even as production practices evolve.

During production, effects, including producing's affect, are applied judiciously to impose upon a recorded sound (or an acoustic sound to be recorded) patterns with specific attributes. Artificial reverberation, for example, superimposes a pattern of reflections on the dry, unprocessed signal to create the illusion of a sound source situated in particular reverberant space. There are patterns to it. Listeners detect these patterns and sense the illusion of space. Since the choice of features is significant, the producer's expertise is manifest while they are detecting, appropriating and replicating salient patterns, whatever these patterns so worthy of attention may be. The producer's attention zeros in on some types of sounds.

3.1.1   Patterns and Representations

Rarely in a recording session are any patterns other than musical patterns called "patterns" or explicitly described. But in order to imitate or recreate a sound (whether previously recorded or natural), patterns and parameters must be consciously identified, if not labeled as such, and manipulated. What makes a pattern a pattern? What features are identified and replicated? Music production has not provided many direct answers for these questions nor formalized methods for seeking them; and while producing remains embedded among artists and engineers who make their own types of patterns, it is not easy to see the producer's direct contributions to pattern making and the final product. However, similar questions are being investigated in related fields of study. Existing literature on the theoretical and computational modeling of sound and music, music theory and music cognition offers much for defining producing expertise; and also provides precedence for modeling producers' creative processes. The following sections consider this literature, moving from synthesizing patterns to generating musical patterns in particular to the performance of musical patterns.

In sound synthesis and computer music composition (that is algorithmically controlled or assisted composition), analysis for the purposes of synthesis is common. The pattern of an instrument's timbre or the procedures of composing are deconstructed and represented formulaically. With procedural representations (e.g. add these frequencies together, move from this chord to that one), patterns become repeatable and also easy to adapt. If it can be identified, it can be acted on with intention. To make these formula or algorithm-like representations, sound synthesis concerns itself, though not exclusively, with modeling real acoustic instruments and computer music with compositional and performance techniques that resemble the psychophysical, cognitive and biomechanical processes of human composers and performers. In these disciplines, real-world archetypes provide templates for computational processes that render patterns as well as criteria by which to evaluate the (success of the) output of sound and music-making systems. A synthetic trumpet is compared to a real trumpet, but also, fantastic timbres are compared to realistic timbres along some criteria. There are expectations and conventions that dictate what realistic trumpets sound like, just as there are expectations, conventions and musical rules that stipulate chord patterns in pop or jazz music. Musical genres and styles are associated with certain rhythms, typical melodic and harmonic patterns, etc.; and these associations are often formalized in theories of music, for example, the theory of Western tonal harmony.

Computer music composition has made extensive use of theories of Western tonal harmony as a basis for designing composing algorithms. These theories represent musical patterns in ways that owe much to (Western) music notation, which uses the parameters of pitch and time to represent sonic events. In Western music notation, dynamic or technique instructions are typically incorporated only intermittently, weighting the importance of pitch and rhythm over other aspects of musical performance. This form of pattern representation offers limited guidance for the actual performance of each note indicated in the pattern. Notation and theories represent potential without the meaning of a particular sound genuinely occurring in a specific context as performed by a specific performer. The sound of genuine occurrences also contains discernable patterns; and these are equally essential for realizing a performance and integral in the experience of musicality. For example, instruments balance dynamically relative to each other and in relation to a given acoustical space or social event (or both, such as in a church wedding procession that is celebratory while demure), and more importantly, performers add expressive gestures. Genuine occurrences are meaningful. The meaning of a pattern is tied to its performance in a particular situation, as a response to that situation.

In computer music systems that generate audible music autonomously, rules may be created (programmed) to describe—to represent the expertise of—how musical and performance parameters change independently or in relation to each other in genuine occurrences. By combining distinguishable types of expertise to construct musical patterns and the techniques used to perform them, the systems simulate reacting to contexts. Thereby they represent (embody) a process of creating experiences meaningful to listeners.

Producing (by humans) similarly constructs occurrences that sound genuine. A music recording is comprised of musical content and performances, and also, a performance situation and environment. Producers act on strictly musical patterns (those that could be notated or described by a music theory), and also, the attributes of performances, recorded attributes and the interrelation of all these parts. Producers do not only shape responses to a performance situation and environment, they operate on the acoustical features of that space, the space in which the recorded instruments (appear to) exist, which with technical mediation may or may not resemble spaces that occur naturally in the real, un-recorded world. A recorded/constructed space may also be evocative of a socially or otherwise demarcated circumstance. One of the things that makes recording fascinating, as an art form, is the way it captures and/or constructs specific occurrences of sound, specific performances in specific contexts.

3.1.2   Complexity

In the studio, one way to construct specific occurrences is by performing with technology. Engineering performances can enhance musical performances. Producers attend to both the engineering and musical performances and how they interoperate. This adds another layer of complexity to analyzing, modeling and emulating producing. Then again, humans sound good precisely because they are complicated and wide-ranging; and because they are subtle, efficient, resourceful, inconsistent and artful in their actions. Computers it is worth noting have none of these traits by default. Technology must be programmed to exhibit these behaviors autonomously or to capture the necessary input from a human user to make up the difference. Existing recording technologies, such as automation, provide insight into what producing does to technical performances to create recorded occurrences.

Mixing console automation is software that "replays" fader movements. (Faders are a type of sliding control on mixing consoles used to adjust channel levels.) Console automation is now ubiquitous. Even wholly virtual systems, in which the notion of a physical fader is somewhat incongruous, offer comparable functionality through graphical user interfaces. For decades, engineers adjusted faders on recording consoles manually, and they did so very musically. At least in popular music recording, where it would be unusual to have a conductor, rarely did engineers and producers simply set a balance and leave it. Minor adjustments and accentuations were performed constantly while mixing, sometimes for technical reasons, such as to avoid levels that would cause distortion, but very often as an expressive gesture.

Traditional faders on old mixing consoles afforded musical instrument-like tactile control, flexibility and degrees of freedom. Console automation enables the sound engineer to record fader movements, and then instructs the machine in their re-enactment. Once the automation has been programmed, every time the automation is run, it renders exactly the same fader adjustments. A human cannot do this. A human cannot perform the same task exactly the same way twice (which is a reason why the ability to record specific occurrences is compelling… as is the ability to create algorithms that re-enact human-like processes). By affording additional technical control and new control mechanisms, fader automation has made it possible to construct new types of recorded occurrences. It did so by capturing human-performed movements, and by offering new freedoms to juxtapose and enhance those gestures with exacting care and deliberation. Traditionally, engineers perform fader moves and operate the automation. However, they do so in consultation with the producer, especially if the movements are for aesthetic reasons. The ability to execute the move lies in the hands of the engineer, but the expertise to evaluate the affect of the move lies with the producer.

Designers of virtual control surfaces face the challenge of matching the control, flexibility and degrees of freedom available in tradition consoles while also affording the advantages of digital and virtual interfaces. Knowledge about why and when certain types of movements are desirable and how they will be enacted provides a basis for developing new control surfaces for music mixing. As traditional recording consoles give way to new control surfaces, some physical movements and some aspects of the model will become obsolete, but not those pertaining to the musicality of executing a fader move. The need for musicality persists, and has persisted for a long time. If musicality is not to be found in the technical aspects of executing performances, where then does it reside?

To summarize,

- It is common practice in producing to imitate sounds not only from the un-recorded world but also from other recordings.
- Imitation and mimicry requires deconstruction and analysis, formal or intuitive, of a template/archetype so that salient patterns may be identified.
- Sound synthesis, music theories, computer music and generative music systems all offer approaches for parameterizing and modeling components of the musical experience.
- In modeling musical processes, it is possible to distinguish among musical, performance and acoustic patterns.
- The patterns that emerge within a genuine performance are essential for making meaning in musical experiences.
- Performances happen relative to, in response to, a context and circumstance.
- In music production, performances on musical instruments and on the recording technology are used to construct occurrences, contextualized performances.

The application of console automation is not in itself a production decision. It is a way to enact one. Musical creators have been toying with the idea of automating musical performances for a long time. The twelfth century polymath, al-Jazarī developed clockworks and automata that could perform music (Paz 2010). Before designing a synthetic performer, al-Jazarī had to analyze the control mechanisms of instruments, in other words, how to generate sound with instruments. Then he devised a method to encode "instructions" so that his devices would generate the right notes at the right times. Those instructions have little in common with computer code, which is capable of taking variables as input. Al-Jazarī's systems were in a sense hard coded to generate one single pattern or a very, very limited set of patterns. Like stored console automation moves, al-Jazarī's musical automatons followed the instructions the same way every time. Human input was required only to initiate the system, which was a huge technological advance. But such technologies lack many humanistic elements essential to music making and expected in genuine occurrences. In particular listeners expect musical, performance and acoustical patterns to vary in ways that sound humanistic rather than mechanical and that adjust performances appropriately for specific contexts. Though listeners' expectations remain poorly understood, especially those pertaining to what sounds predictable; it can be observed that there are degrees of predictability. A given variant taken individually or a performance or a recording may or may not be predictable, and humanistic variation is not necessarily an absolute quality but rather happens (or not) in response to a context. Humans are instinctively very good at discerning what sounds humanistic.

Producers cause variation in musical performances and/or choose from among variants of performances, matching variations to performance contexts to construct contextualized occurrences. The process is frequently artificial and often requires the stitching together performances that did not genuinely occur in the context heard on the finished recording. Nevertheless, the end product (ideally) offers listeners an illusion of a performance that occurred genuinely in a believable (if not realistic) context. The process requires producers to make predictions about listener expectations and interpretations of what is heard.

3.1.3 Generating Musical Structure

(Human) composers know a great deal about generating variation. Composers do not compose one piece but many of similar and different types and styles, and through them all the unique styles of individual composers show through. What produces variation in composing? Some composers have experimented with aleatoric procedures to introduce variation in the composing process. During the mid to late 18th century, numerous musical games were published that used dice to introduce variation into compositions. By casting die an amateur musician could "compose" varying compositions. (Some of these games have been attributed to Mozart.) The games consisted of numbered musical segments and rules. The rules instructed players to combine and recombine the segments such that it produced a coherent piece. (Hedges 1978) In the 20th century, taking this idea to a conceptual extreme, John Cage introduced "chance music", injecting aleatory into the most fundamental structures of his musical compositions, which also served to open up the range of sounds or sonic events that may constitute a composition.

After Cage, composers continued to develop techniques that instructed performers in processes and procedures for playing patterns, further dismantling established systems for specifying which notes to play when. These experimental processes yielded distinctive textures, for example, in the minimalist works of composers like Terry Riley, Steve Reich and La Monte Young and Philip Glass. Between 1967–1969, Cage and composer-scientist Lejaren Hiller collaborated on a computer music composition, (Cage 1967–1969) for which Cage created "chance operations". In 1968, of the creative potential of computation Cage observed, "there are various ways of producing [creating] music, and writing is one of them…'making' is certainly another." (Austin 1992, p. 15) In the 21st century, musician Beck continued riffing on these ideas with his "Song Reader", a collection of twenty pieces published elaborately as sheet music (Hansen 2012) and interjecting the recording process, he used an accompanying songreader.net website (Hansen 2013) to encourage fans to interpret and produce versions of these songs.

Additionally, there are theories of music that describe less variable processes associated with conceiving of musical patterns. Human musicians are limited in what they can create by knowledge, technique, perception and/or circumstance. When they create, mimic or imitate sounds, music and performances they can do so only along a finite number of parameters (for any given occurrence), making it similar to the model/template in some, perhaps many, but not all ways. In their Generative Theory of Tonal Harmony (GTTH), Lerdahl and Jackendoff assert that experienced listeners have preferred musical interpretations, and this provides reasons for the (choice of) parameters and constraints their generative theory stipulates. Consistent with their explanation of preferences, the GTTH pertains to patterns that may be described in terms of hierarchical relationships: grouping structure (of motives, phrases and sections); metrical structure (of strong and weak beats), time-span reduction (the importance of pitches within grouping and metrical structures); and prolongation reduction (which places pitches in a hierarchy along the lines of harmonic and melodic tension). (Lerdahl 1996) This is one way to describe how musical structure is conceived and perceived. Other descriptions are possible. But in many of the generative models of music formulated thus far, Western tonal harmony or another formal theory for describing musical structure provides a basis for determining the parameters for analysis; and thereby set aside many performance related attributes and other features that arise in musical practice. (For introductions to topics related to generative models see also Rhythm: (Longuet-Higgins 1984; Repp 1990); Melody: (Narmour 1990); Chord-sequences: (Johnson-Laird 1991); Jazz: (Steedman 1984); Machine Models: (Schwanauer 1993); Music and AI see (Balaban 1992); Probability: (Temperley 2006); Algorithmic composition in live performance: (Collins 2003).

Theories are by necessity generic. They cover a range of variants that share commonalities on some levels. The rules of Western tonal harmony apply to many styles and genres of music. Starting with Western tonal harmony, adding additional constraints, it is possible to imitate (parameterize and emulate) specific styles and composers. Meyer offers the following definition of style: "Style is a replication of patterning, that results from a series of choices made within some

set of constraints." (Meyer 1989, p. 3) Creators impose constraints idiosyncrati-
cally, but also pattern generation is constrained by the degrees of freedom of the
medium, resources and instrumentation, and the creator's understanding of the
goals. David Cope, a composer, develops algorithms that generate new works in
the styles of historical figures in music. His "Experiments in Musical Intelligence"
(Cope 1992, 1996, 2004) resulted in computational systems that analyzed corpora
of existing compositions by well-known (Western) composers, extracted patterns
that characterized those composers' styles, established margins of variation for
those patterns, and then rendered new compositions in these styles by making
new combinations of patterns and synthesizing similar patterns. (Cope 1996) For
Cope, styles are revealed in a composer's "signature". "Signatures are contiguous
note patterns which recur in two or more works of a single composer and therefore
indicate aspects of that composer's musical style. Signatures are typically two to
five beats (four to ten melodic notes) in length and usually consist of composites
of melody, harmony and rhythm. Signatures typically occur between four and ten
times in any given work." (Cope 2004, p. 109) Signatures that appear frequently
throughout the composer's body of work appear in different contexts and guises.
But, signature patterns alone are not enough to generate characteristic musical
structures. Signatures are woven into structural contexts that also exhibit identifying and identifiable patterns. Cope recognized "earmarks" to demarcate positions
and features of structural significance within a given composition. (Cope 2004,
p. 114) (For more on the computational modelling of musical style Dubnov in this
volume).

Relatively infrequent or idiosyncratic patterns operate within a context of less
idiosyncratic, familiar, recognizable musical material. Some material is familiar
because creators are exposed to music before they begin creating it, just as listeners have listened previously. Both music creators and listeners bring knowledge of
music theory, style, convention, etc. to the creation/listening experience. Musical
compositions are constructed of and (analytically and cognitively) deconstructed
into patterns in harmony, rhythm, form, expectation/realization, consonance/dis-
sonance, theme and variation, etc. in part because acculturation lets the creator/
listener know this is the material of music.

Moorefield has likened record production to composing (Moorefield 2005)
highlighting the many ways that building structure, stating motives and themes,
presenting variation, etc. in a recording is similar to traditional composition. If
this is the case and if Cage intentionally incorporated musical and non-musical sounds into compositions using chance, why cannot producers deliberately
produce variations and then organize them with compositional intention? But
these composition-oriented ways of defining structure do not capture all that is
heard; nor describe much of the material of record production: patterns pertaining to sound quality, balance, depth, feel, groove, punch, edginess, soul, etc.
On recordings, these types of patterns work in tandem with those more closely
associated with composition. Both producing and composition aim to produce
coherent, structured listening experiences utilizing what listeners know and can
detect.

3.1.4 Performances

Composers indicate which musical structures are to be rendered by performers and when. They use tempo and dynamic markings and denote special playing techniques, but they do not, as a conductor might, direct the performance of
those instructions. Unlike (traditional) composers, producers have direct influence
over musical performances; but in the studio where performances are recorded,
re-recorded, edited and combined, the line between influencing a performance and
creating one can be rather indistinct. Albin points out, "[A]n increasing number
of artists have adopted synergistic creative approaches where songwriting, arranging, and recording proceed simultaneously. Peter Gabriel is an example of an artist
whose songs are conceived and developed as recordings from the start...[w]orking
always with a tape recorder, he deals with recorded sound from the earliest sketch
stages and is guided throughout the compositional process by his responses to aural
images." (Albin 2001, p. 30) Something is happening during the recording process,
as a result of performances being rendered on a recording medium, rendered into
a recorded context and played back that provokes compositional and performance
responses. Producers respond to these recorded attributes, and they influence live in
the studio performances to construct recorded ones.

Both the control afforded by repetition and the ostensible permanence of the
recorded document encourage precision in pitch, rhythm and synchronization, etc.
They also bring forward the far less quantifiable though potentially far more mean-
ingfully aspects of the performance. Keil, a cultural anthropologist, observes, "It is
the little discrepancies within a jazz drummer's beat, between bass and drums, between rhythm section and soloists, that create 'swing' and invite us to participate."
(Keil 1987, p. 277) To allow for meaningful participation, by both performers and
listeners "[m]usic, to be personally involving and socially valuable, must be 'out of
time' and 'out of tune.'" (Keil 1987, p. 275) (Slightly) out of time and out of tune
patterns pull and push. They communicate well because they diverge (or do not
diverge) from the prototypical template participants understand, and this has meaning. This invites appraisals, inferences and interpretations. There is in the recorded
document and inherent tension between precision and discrepancy.

What do these discrepancies mean? To understand, to imitate or to emulate the
pull and the push of individuals or styles requires models of pattern making based
on theories of music and attributes associated with common performance techniques. Such models need to represent practices of performing music, such as the
rhythmic discrepancies that lead to groove. Iyer (2002), for example, has investigated the influence of embodiment on expressive timing. The performer performs
using a body (with physical constraints and affordances and conditioned patterns).
The performer's cognizing is situated. Performances are performed from within an
environmental and social context. Iyer analyzed renowned jazz recordings seeking evidence for these influences in micro-rhythms. In a recording studio, where
musicians often record performances at different times (without sharing an environmental or social context), layering through overdubbing rather than playing together, getting a groove may depend on the producer's ability to both influence

individual performances and blend the performances. Modeling a producer's style may require understanding the producer's influence on performance practices and/ or the patterns left by a producer's performances on instruments or on technology. During a recording session, the producer's actions cause some sonic features to be emphasized over others or some types of patterns, affects or variations to occur. Discrepancies are not random. There is pattern to them, even if the causes are not always known.

3.1.5   The Patterns of Producing

There are discrepancies in recorded performances, and among recordings there are discrepancies of another sort. Recordings differ because the elements along the (audio) signal chain from the performer and the room to the recording devices (and post-production processes) are different and are configured differently. This results in variations in, for example, sound quality. Because there are many attributes that factor into listeners' perceptions, sound quality cannot be assessed simply in terms of good/bad or how good/how bad. Different attributes contribute differently to appraisals and contribute differently at different times to different appraisals. Thus, there is a fundamental problem of identifying which qualities best represent a sound recording (or sound recordings in general), in other words, choosing the parameters for a model or for evaluating a template. In practice, producers and engineers land on an appropriate set of parameters intuitively. In research, there are numerous approaches to sound quality assessment and varied criteria both subjective and objective have been used in evaluations. (For an introduction to the study of sound quality see (Letowski 1989); and from a psychoacoustic perspective see (Fastl 2005)).

Production decisions can cause discrepancies in sound quality. When mimicking the sound of or a particular sound in referenced recording, producers emulate discrepancies produced by other producers, engineers or artists. Some features are emulated so often and the production style becomes so recognizable and well known that it becomes a normative template against which new discrepancies stand out, for example, sounds associated with a region such as the Philadelphia or Seattle sound. Other styles are so idiosyncratic that they remain tightly associated with their originator, for example, Phil Spector's "Wall of Sound".

As the work of Lerdahl, Jackendoff, Cope and others has shown for music composition, in modeling a processes or emulating a style, choosing which idiosyncratic patterns or discrepancies are the meaningful patterns requires knowledge about—or taking a decision about—the creator's frame of reference and his[1] preferred interpretations. When modeling Western musical styles, because of it is prevalence, it is feasible to make some assumptions about knowledge that is shared among composers and analyzers. Lerdahl and Jackendoff made thoroughly considered assumptions about what *educated* listeners are likely to prefer based

---

[1] "He" is used through out the chapter because the vast majority of producers at the time of writing are male. (Hopefully, this will change.).

on what they are likely to know and perceive. But, it is not a given that creators and observers do share knowledge about musical structure, the context, constraints or degrees of freedom. (Lefford 2004). In recording, particularly popular music recording, these assumptions are harder to make. There are no prevalent, widely studied rules of music production (although listening audiences are becoming more literate about production). Without this basis, if the producer wants the original/ template/archetype to be recognizable in the emulation, if he wants to make a reference, he needs to consider what patterns the listening audience finds salient in both the original and the emulation. If more straightforwardly, he is interested in aligning the audience's appraisals of this emulation with his own, he will consider what features are likely to be detected and likely to command attention regardless of the knowledge of the reference listeners bring.

To summarize,

- Performances, compositions and recordings vary.
- At some of the lowest micro and highest macro levels, variations might be represented by aleatoric operations. Nevertheless, other patterns are more consistent in nature.
- Creators' idiosyncratic processes leave identifying signatures. However, what constitutes an identifying signature remains a matter for ongoing research as does predicting what combinations characteristics and idiosyncratic patterns will be successful in emulations.
- Listeners/creators have expectations about variation and consistency, and this presumes that creator/listener bring some a priori knowledge to the acts of creating/listening. This knowledge impacts the creator's/listener's understanding of the context, and therefore the meaning of certain patterns within.
- Discrepancies from an ideal template—whether stylistically idiosyncratic or resulting from other constraints— is what invites participation and meaning making.
- Producing encourages variation and discrepancies, and performing as a producer also creates characteristic discrepancies.

3.2   Producing

Recordings are rarely made for those who participate in recording sessions. They are made so those who are not present, who are not privy to the creative process or do not have direct access to the artist may listen and participate.

3.2.1   For the Masses

Producers aim to produce recordings that will be listened to by many, preferably repeatedly and moreover purchased. To that end, producers make assumptions about what the audience knows and is able to detect, and applies influence to performances to engage those listeners. If the goal is the broadest audience possible, in popular music production at least, that means expecting to share little knowledge of music theory, style or convention with the audience. This can be contrasted with classical music production, in which producers may expect not only knowledge of Western tonal harmony, but also, of the classical repertoire. In other idioms, producers might or might not assume genre specific knowledge. At the very least, regardless of genre, producers can expect listeners to share some common knowledge that stems from experience listening to music and other sounds (on and off recordings), about sound and music in social and cultural contexts, about interactions through sound, and not least of all, which is necessary for detecting genuineness and discrepancies in performances. This general knowledge enables listeners to participate in meaningful exchanges using sound when the sound is not recorded and also to experience meaning in recorded sound. The activation of these participatory skills while listening to recordings mirrors social activity.

Accordingly, to engage audiences, producers have available features associated with initiating and sustaining social interaction with listeners. What types of sonic features are these? Much about auditory perception stems from facilitating social interaction. Our perceptual systems evolved over millions of years to offer advantages that improved the odds of survival. Humans hear and attend to things and in ways that have benefited us as a species. These traits, arising from natural selection, are common among all hearing humans. This concept is fundamental in understanding the ontogenesis of our sensory organs. As human hearing evolved, our oral capabilities concurrently evolved to enable more complex participation in more sophisticated social interactions. Humans are inherently social creatures. Our survival as a species depends on socializing. The survival of individuals depends on socializing.

The next several sections develop what is in essence a socio-biological view of producing by pooling concepts from perception, cognition, psychology, biology, linguistics and musicology. Music research has already produced several socio-biological interpretations of music-making and evolutionary perspectives on it, but their direct applicability to sound recording and music production remains largely unexplored. (For varied introductions to the topic of music and evolution see (Wallin 2001; Cross 2001, 2003; Dissanayake 2008; Fitch 2006; Hargreaves 2005; Laurel 2006; McDermott 2005; Justus 2005; Patel 2006)). Perception and biology, because their processes and phenomena have been scrupulously investigated, offer insight into additional ways to parameterize produced music, thereby making it more readily available for analysis, modeling (cognitive, computational or otherwise) and ultimately for sophisticated, technological intervention. These investigations start close to music and move steadily toward comparisons with socio-biological processes found outside the music domain.

3.2.2 Creating Social Experiences on Record: Connecting with Listeners

To produce is to cause. Producers cause connections between recording artists and audiences by influencing and/or directly manipulating the qualities (as well as quality) of the performances, musical structure and acoustic features of recorded sound, and in popular music production, through the song. Songs provide both content and structure for performances and impact how performances are contextualized, appraised and understood. At the center of a song is a voice. About "songfulness", Kramer offers that there is in it "a sense of intimate contact between the listener and the subject behind the voice. This contact is both an aesthetic relationship (that is, an embodied fiction) and, perhaps, an indication of the specific fantasy-structure that underlies the experience of songfulness." (Kramer 2001, p. 53) That subject behind the voice has character. The fantasy structure is a set of relationships and some extend to the listener. Deconstructing produced vocal performances reveals much about how connections with listeners are made, what structures support the listening experience, and how fictions are embodied.

Every hearing human is intimately familiar with the sound of the voice. The voice has fluid fluctuating qualities that are difficult to qualify or quantify, but voices devoid of variation appear artificial. There is meaning in the discrepancies, and the misinterpretation of them has been source of conflict throughout human history. Of the album *Pet Sounds* (by the Beach Boys, produced by Brian Wilson (1996), Moorefield writes, "The complete blend of voices, the utter precision of the intonation, and the exquisite ensemble of attack, phrasing, and expression are everywhere evident. One cannot help but wonder just how such extraordinary craftsmanship and aesthetic beauty was achieved." (Moorefield 2005, p. 17) Moorefield's concise digest covers numerous sonic parameters, the layers of information made available to listeners through the recorded performances: pitch, harmony, timbre and rhythm and each with macro and micro musical structures. He alludes to localized events and patterns that unfold over timeframes of varying lengths. However, even though voices singing words occupy a primary (and potentially the primary) channel for communication in this recording, the lyrics are not mentioned in his appraisal.

The language obviously holds semantic meaning and informs. Also through word choice and usage, language provides listeners with clues about (social and cultural) context and the singer. Yet, listeners of *Pet Sounds* do not infer that it really would be nice from the words "Wouldn't It Be Nice". They infer the nicety from just about everything but those words. Moorefield's experience of the voices as being full of concrete meaning that is not conveyed through lyrics is to be found in all forms of musical expression, but in recorded music, precisely because it is recorded— because discrepancies are so meticulously presented and so reviewable— the effect is especially discernable. Anything listeners learner about the singer, song and context, must be learned from the recorded sound. It is the only source of information. They have no body language, no visual signs. They cannot see what the singer sees. All must be inferred. And, the sound, that exact sound, as it unfolds over time and as it is replayed, can be checked and re-checked for evidence to support appraisals.

The knowledge, expectations and understanding the audience brings to this appraisal may be dissimilar from perspectives of the artists and producers. There is no real-time feedback from a live audience by which to gauge responses. The producer, artist and audience are distanced by time, space, and possibly social and cultural circumstance. The challenge for the producer is to convey information through shared channels of communication; without any agreed upon set of mutually understood references or signs, without the benefit of a grammar or lexicon, but instead by fruitfully exploiting other indicators and cues. As when communicating with children, or even more distantly, in our Doctor Dolittle[2] like discussions with pets, words may not communicate or may not communicate enough. In animal-to-animal communications, Marler finds mismatches where, "the listener's perspective may be fundamentally different" from that of the animal emitting the sound, and therefore the listener may respond not at all, unexpectedly or with negative consequence for the creature uttering the sound. (Marler 1961) via (Seyfarth 2003, p. 147) Yet, experience shows that complex meaning may be indicated vocally and non-linguistically without giving in to every solicitation for a new toy or more food. This requires a shared context (or the assumption that the context is shared).

Perspectives and interests among communicating parties often do differ at least somewhat, but points of mutual interest help to establish connections. Children and (some mammalian) pets, for the sake of survival, are very interested in bolstering connections with caregivers; and children, dogs and cats are creatures that show acute sensitivity to non-linguistic vocal signs and cues. Children possess a wealth of experience in making themselves understood, although verbal skills have only recently been acquired. The child communicator has only just learned from the parent that, without language, by manipulating patterns in the sound, voices may soothe, discipline, encourage, reassure, indicate, affirm, play and engage[3]. It is no wonder that they are quick to accept fantastic stories about talking animals and interspecies communication. Children know better than adults that a common language is not essential for communicating through sound.

Through recorded sound, producers are able to sustain nuanced, meaningful exchanges with unknown individuals without the benefit of a common (musical or linguistic) language. What vocal features are available to the producer and why do they communicate? The following deconstruction of producing investigates

- Vocal features in recorded music that engage
- The affect of these features (on audiences)
- Reasons for the effectiveness of these features
- How these features are enhanced by the influence of the producer.

3.2.3   Listening to Voices

Humans have specialized physiological mechanisms to support the capacities to perceive and vocalize speech utterances as well as other non-linguistic indicators and cues. We perceive in voices enormous detail. This aptitude parallels fine motor

---

[2] A children's book character, Doctor Dolittle is a fictitious veterinarian with a preternatural ability to speak with and understand animals. (Lofting 1920).

[3] For many indicators, signs and cues, the learning process involves correlating auditory patterns with visual patterns or other indicators. Regardless, the patterns of non-linguistic vocal communication are established.

control over the sundry components of the vocal apparatus, the larynx, lips, tongue, velum and jaw. (Fitch 2000) Human's enhanced respiratory control may also factor in the complexity of our vocal communication, particularly in speech. (MacLarnon 2004) Specialized cognitive processing supports this physiology. "Vocal expression may be construed as an evolved psychological mechanism that serves crucial functions in social interaction." (Scherer 1995, 2003, p. 73)

Non-linguistic vocalizations are as integral to vocal communication as speech, whether directed toward pets, children or adults; and accordingly, as illustrated by the *Pet Sounds* example, lyrics are not the most informative part of vocal performances. Non-linguistic communication in humans and vocal communication in other animals has been widely studied. (See Bradbury 2011) for an introduction to animal communication and (Wiley 1983) for an introduction to communication theory as applied to animal vocalizations.) Interesting, non-linguistic and vocal communication as a field of study came into its own in the 1940s when the cost of recording technology equipment, audio and film, fell within the means of scientists. (Kendon 2004, p. 69)

In every human activity involving vocal communication, be it spoken or sung, non-linguistic vocalizations divulge information about the person making the utterance, about gender, age, level of activity and state of health. From the sound of the voice, the timbre, energy, articulation, dynamics, pitch, rate, prosody and rhythm (to name just a few parameters), listeners make inferences. Conversation, dialog, instructions, explanations and sung verse are all coupled to non-linguistic indicators that are as efficient, if not more efficient, than language for expressing meaning.

Recorded performances convey information that allow listeners to infer the physiological state and/or social circumstance of the singer (e.g., tired, excited, (dis) empowered, commanding of authority, open/withdrawn, etc.). Even the most naïve infant recognizes that, in no small way, emotional state, the valence of the emotion and the level of arousal (i.e., how much energy is behind the emotional expression), is communicated through non-linguistic channels. In all sorts of contexts, listeners rely on non-linguistic information to interpret intentions and actions, and to assess potential threats and opportunities, making it integral to coordinated activity.

The same types of indicators and cues are heard in recorded vocal music. Combined with sung lyrics and musical patterns, they provide listeners with information: Who is the singer? (Wadhams 2001) Who is the singer in relation to the song? What is the singer feeling and as a response to what? What is the singer likely to do in the given circumstance the song establishes? What is this song about? (Wadhams 2001) Inconveniently, on recordings as in life, indicators and their meaning are not always perceptible or understandable. The producer is expected to supply perspective on how listeners are likely to appraise, on the reception of, what performances communicate; and act to improve that communication.

3.2.4 Perceiving Possibilities

Among listeners, as perceivers, record producers are somewhat unique. A great deal of the producer's job entails critically assessing performances, envisaging improvements and creating processes to actualize the imagined. Along the way, the producer prioritizes certain sonic features and elements of performances, acting on and affecting them to make them better. Defining "better" is also part of the producer's job (and a would-be emulator needs to consider the producer's preferred interpretation because it guides his actions.) The criteria for better may relate to any number of attributes such as technical execution, emotion, tone, etc. Different criteria apply under different circumstances; but whatever the desired attributes, they lend themselves to improvement only if they are detectable and comprehendible to the producer and to an audience making appraisals. The skill required to identify what is needed for improvement—not unique to producers, but over-developed in producers—is interesting because it requires actively seeking patterns that are not as perceptually salient as they might be, patterns that do not stand out as clear patterns relative to their context; and this runs askance to accepted notions of how perception generally works. Listeners are expected to gravitate toward salient patterns. Producers gravitate toward and act on noisy signals.[4]

One common action taken by producers comes in the form of suggestions to artists regarding performances. In historical video footage of a recording session for U2's *Unforgettable Fire* album (1984), lead singer Bono records vocals for *Pride (in the Name of Love)* with unbridled passion. After the take, producer Brian Eno comments, "The only direction I might offer you is that the first chorus might be a little more restrained than the others, but I wouldn't like to inhibit what you're doing." (Video appears in (U2, Guggenheim 2011)

Here again, in the evaluation of the performance, no reference to specific lyrics or their performance is made. Eno's comments operate on a structural level. By showing restraint early on, the performance of the second chorus relative to the context of the entire recording is by contrast more intense, and indicates change to the listener. That additional information means something. By contriving a change, a more meaningful performance is constructed. The semantic meaning alone does not tell the whole story here, or for that matter in most recordings. It is worth noting that many listeners listen to music in languages they do not speak, and articulation in recorded performances is at times lax particularly in popular music. That is not to say gibberish is sufficient. The importance of lyrics should not be deemphasized, but apportioned within the context of the greater whole.

Wittgenstein's concept of a "language game" (Wittgenstein 1953) illustrates just how language usage, meaning and context are all interconnected. Words or phrases derive their meaning from the context and how they are used. Language usage is like a game with rules, constraints and goals such as to convey intended

---

[4] Information Theory (IT) (Shannon 1949) has been applied, both loosely and strictly, to innumerable areas of study. With relevance here, Kendon has applied information theory to interaction models and behavior (Kendon 1990, p. 25); (Baddeley 2008) offers a review of Information Theory and the brain, perception and behavior; (Cohen 1962) offers a foundational critique of the application of IT to music, also (Meyer 1957). Exactly how IT is applied to music and aesthetics is an area of ongoing research and debate. For the purposes of this discussion, information is being communicated between a sender and a receiver. That information may be more or less discernable. Of interest is the clarity. How information theory specifically applies is beyond the scope of this study.

and intelligible meaning. Also like a game, where strategies are employed under different circumstance to variable effect, language usage affords degrees of freedom. (i.e., "He hammered away at the guitar's fingerboard." possibly refers to causing physical damage to an instrument or an intense practice session.) Depending on the intended meaning and the context, different strategies may be applied to greater or lesser affect. The "game" is the same whether the words are spoken, written or sung, but in sung performances, the context includes lyrical, musical and acoustical content.

3.2.5 Affectations and Artifice

Shifts in affects result in shifts in meaning. Shifts in contexts result shifts of affects and thus meanings. Affects, whether greater or lesser, are influences usually of a psychological or emotional nature or that produce experiences of a psychological or emotional nature in those who perceive them. As a description of resulting experiences, the word "affect" is useful, concise and problematic. It lends itself to interpretation depending on the context of its use, but the same may be said of any pattern embedded in music communication. In daily conversation, affectation is associated with contrivance. In psychology, affect relates to emotion; and cognition, behavior and emotion are viewed as distinguishable phenomena. Affective science deals with experiences and displays of emotions and the recognition of displays of emotions in others (including by computers). In basic scientific terms, affectations are behaviors or displays, in this case vocal behaviors and displays, with distinctive sonic qualities. There are several theories about how these qualities may be characterized, especially as they pertain to emotional cues. One idea is that affectation are associated with particular (sets of) audible indicators, and listeners detect discrete cues. (Juslin 2008) which also provides a review of the literature on vocal affect and an explanation of the biomechanics and physics involved. Bachorowski and Owren refer to this as a "cue-configuration" theory. Alternatively, listeners may be detecting changes or relationships among "underlying dimensions". (Bachorowski 2003) This paper offers an interesting discussion regarding the different theories.) For the purposes of this discussion, the term *affection* is used to describe vocal qualities that impart non-semantic information, whether it is communicated via a non-linguistic vocalization or an affected linguistic vocalization, and whether the content belays emotion or other types of information.

Vocal expression is one channel through which emotion is communicated, and emotion is a constituent part of the musical experience. Music of all forms displays emotion and triggers emotional experiences, but it is not the experience in total. Hargreaves and North delineate three psychological functions of music: cognitive, emotional and social. (Hargreaves 1997) The term "affectation" connotes that emotion and behavior are connected. Listeners at least instinctually know this. The ability to associate certain behaviors with certain emotions (and sounds) is a social skill and evolved adaptation. The term also highlights that there is an element of intentional if not outright, deliberate contrivance in musical performances. They are acts, behaviors that are controlled and displayed with intention and for intended meaning. Some of that meaning usually pertains to emotion; and the resulting sound is intended to be affecting to those who hear it. Producing intentionally exploits listeners' abilities to detect vocal affectations that communicate information. Producers are experts in how affectations are performed and the affect of certain affectations on listeners.

Contrived though they may be, recorded vocal performances are not necessarily disingenuous. They may be representative, though exaggerated or premeditated, of feelings really felt and states really experienced. Performances and singers (and producers and recordings) are judged not only for their character and technical quality, but also, for their authenticity. This pickiness seems slightly mismatched with our modern context. It is easy to understand why we would be highly discerning in earlier centuries, but today, we tolerate innumerable daily encounters with voices that are in fact not real but representations of voices captured in media or synthesized with electronic devices. Maybe our preferences will change over time, but regardless, in the meantime, the conversion of acoustic signals into electronic ones leaves audible artifacts that can distort the nuance and subtlety of live vocal affectation. Storing them on a medium or transmitting them also effect the sound. These artifacts can impart unreal qualities. For one, sonic information is lost during processing. Listeners listen for these artifacts.

Experts in sound quality will argue in what ways particular losses are or are not detectable, but of interest to producers is the fundamental potential to influence and shape the qualities heard in audio signals intentionally. Mixing console automation is one example of a technology that adds information (about dynamics) to recorded performances, but the variety of signal processors available is vast as are the possibilities for generating and layering all manner of performance, musical and acoustic information onto recorded sounds. Producing overtly capitalizes on the inherent artifice of the recording medium to manipulate performances, to heighten the presence and/or clarity of the vocal exposition, to produce desirable affectations and exceptional, alluring singing performances.

To improve sound quality, noisy signals may be cleaned. The musical attributes of pitch, harmony and rhythm may be perfected. Current technologies and recording techniques offer producers and sound engineers exacting control over many of the technical aspects of recorded vocal performances, so much control that what should be considered too much exactitude is frequently debated in public forums along with the value of the technologies that afford it, such as the Antares "auto-tune" plug-in (Antares 2012). In 2009, producer Rick Rubin acknowledged, "Right now, if you listen to pop, everything is in perfect pitch, perfect time and perfect tune. That's how ubiquitous Auto-Tune is." (Tyrangiel 2009). (See (Hamlen 1991) for a discussion about singing ability, perceived quality and commercial success.) But the application of these tools is not what makes a recording sound produced. Recordings made without these tools sound produced, even to modern ears. Although the fidelity and accuracy of the signal itself is essential to consider, listeners are more likely to attend to clear signals, these attributes alone are not enough to produce a noteworthy recording,

3.2.6   Getting the Message Across

Among non-expert listeners, the performance characteristics most often remarked
upon, for example in the music press, are emotional power and tenor, but unlike
the more technical attributes of performance like pitch, these are much more difficult to compare (parameterize) levels of emotional display among performances.
Every display must first be related to its unique context. An intense response to one
context may not sound intense in response to another context. Indicators of emotion
isolated from context are recognizable (i.e. that is a sad/happy voice), but emotions
like words are only meaningful within a context, as connected to a particular person
emoting under particular circumstances in a genuine occurrence (e.g., a sad sounding voice might indicate sad-lonely or sad-disappointed depending on the context).
On recordings, by combining layers of musical information (melodic and rhythmic
patterns, dynamic structure, etc.), acoustic information and affective information
and the semantic meaning from lyrics, a context is established for each pattern that
appears within. If the components amalgamate in some way, even though this performance scenario has been contrived, they convey a coherent message. Listeners
expect songs about clowns fittingly to have up-tempo, bouncy, child-like melodies.
A song called *Tears of Clown* (1967) made famous by Smokey Robinson and the
Miracles' 1967 recording on Motown needs to communicate a persistent darkness
beneath the playful show, not only with the word "tears" but also through a tapestry
of other signs, sad-sad-sad.

There needs to be multiple statements of sadness. One way to link musical,
performance and acoustic patterns together and to optimize the chances of
comprehendible communication is through redundancy, by delivering important
information via more than one channel. For example, a listener will reliably
detect the parallels among despondent lyrics, a minor mode melody and sad
vocal affectation. A cornerstone concept in perception, the Brunswik lens model,
represents perception as a process of interpretation; utilizing multiple cues, often
conveying redundant information, and knowledge about the environment or context.
(Brunswik 1956) Juslin and Laukka have used the lens model to explain how
emotion is communicated through music (in general, not in recorded or vocal music
specifically). "[E]ncoders express specific emotions using a large set of cues… the
cues are probabilistic in the sense that they are not perfectly reliable indicators of
the expressed emotion." (Juslin 2003, p. 801) also (Juslin 2008).

It follows that indicators congruent with or plausible within the context are more
likely reliable. Hammond and Stewart (Hammond 2001) describe the Brunswick
model as "probabilistic functionalism", functional in the sense that an "organism's behavior is somehow organized with reference to some end, or objective."
(Hammond 2001, p. 3) A sad clown makes sense if there are observable or known
causes or benefits for both the playful and melancholy displays in turn. Multiple
indicators or cues may be embedded within a single (vocal) performance and/or in
multiple channels, multiple performances on multiple instruments, throughout the
arrangement and orchestration. The associations that are triggered are re-enforced
and clarified by the redundancy; by the connections among indicators and between

indicators and the context; and/or by a perceived goal. With these tools, the producer communicates artistic intent.

To summarize, to connect with the listening audience, producers and producing attempts to modulate the singer's affections.

- Humans have perceptual acuities and are conditioned to utilize non-linguistic vocal indicators in social interaction.
- Vocal affections ideally provide listeners with indicators signifying the singer's state (emotional, motivational and physical).
- On vocal recordings, where there are no visual cues, non-linguistic indicators are an integral part of establishing communication between the singer and the audience.
- Listeners appraise affectations through the frame of how they understand the context.
- Where the indicators supplied by a performance are not clear, producing works to improve communication. Recording offers recordists exacting control over the qualities of recorded sounds, which producers deliberately manipulate to make meaningful listening experiences.
- Redundant information delivered via multiple channels improves the odds that signs, cues or indicators will be successfully communicated to listeners.

The goals of record production and mechanisms of auditory perception are intertwined. To understand how the two corroborate, it is necessary to consider in greater detail the impact of recording technology on the audio signal and the acoustic features before they reaches the ears; what singers do in front of microphones to create affectations; and also, the producer's role in crafting both the sung and technologically mediated sounds.

3.2.7 *Producing Vocals Tracks*

Recording technology, microphones, amplifiers, signal processors, even the acoustics of the studio environment all color the vocal timbre to greater or lesser degree and with intended and/or unintended consequences. For example, without the microphone, the crooning style of Bing Crosby would not be possible (Albin 2001; Frith 2004). To croon, the vocal apparatus, like a brass or woodwind instrument, holds a precise configuration while air is pushed with controlled force out of the lungs. The tone is created and constrained by physiology, biology and biomechanics. It is not possible to croon very loudly or very softly. Therefore, crooning over a live orchestra necessitates a microphone and amplification. Microphones enable this and other distinctive vocal affectations found in recordings.

Bing Crosby and Frank Sinatra, singers renowned for their unique, signature styles, developed innovative performance techniques through idiosyncratic musical interpretations, and also, their use of technology—both learned to play the microphone as though it were an instrument. Intentionally moving closer and further away from the microphone's diaphragm and changing axis throughout the performance, they deliberately modified the timbre of their voices slightly, and in this way, created (new) affectations and added (new) meanings to the popular singing vernacular. Other distinctive styles of affectation and technologically enhanced vocal qualities are associated with other genres and singers. Because recording and sound reinforcement are now so prevalent, most singers develop some sort of microphone technique. Singers choose microphones just as instrumentalists choose instruments, for their timbre and feel, for how they respond and make the voice sound. Producers and sound engineers, choose microphones for similar reasons, but do so with the intention of emphasizing certain features of the singer's voice or introducing certain tonal colors for a particular recording.

Production's influence extends beyond vocal timbre and technique to the lyrical structure, which the producer can impact even without advocating for modifications to the words. In vocal music, narrative structure tends to be the rule rather than the exception. In an analysis of popular vocal music of the 1960s, Toft states, "recordists strive to heighten the impact of the lyrics, melody and harmony, through an emotionally persuasive performance designed to enhance the expressive flow of the narrative (Toft 2010, p. 71)". In the Bono example used earlier, Eno focused attention on the build to the songs emotional climax, thereby superimposing form on the exposition, which will ultimately impact how the words themselves are weighted in importance and interpreted. When a producer asks a vocalist to make the performance more emotional, or change the quality of the emotions being expressed, what is he asking the vocalist to adjust, what features are changeable?

The Vocal Instrument The voice, as an acoustic instrument, responds to changes in pressure and tension, to the resonant cavities and openings, to dampening and shape alterations, etc.; all of which discernibly alter the timbre. Some of these changes may effect articulation, rhythmic accents or dynamics. The timing of words, the time between words, and the length of words are other features commonly manipulated. Slurring, spitting and growling each have their sounds, resulting from configurations of these varied parameters. Vocal technique also yields other far less quantifiable though no less detectable colors, for example, aspirations around the words, etc. All these features make up a palette of affectations.

During recording and mixing, producers and sound engineers adjust the timbres of instruments including the voices using transducers, amplifiers and signal-processing devices; each of which provides distinct ways to filter the sound. Equalizers, for example, are used to boost and cut frequencies, which intentionally or unintentionally may impact how articulation or energy is perceived. Artificial reverberation, delays or chorusing signal processing effects might be added to make a voice sound bigger, more substantial or more present. Reverberation places the voice in an acoustic environment, nearby or far away. Dynamics may be limited or created. Any of these techniques alter vocal affections or how they may be understood relative to their perceived context. A shout up close means something different than a shout at a distance. In this way producing alters listeners' appraisals of the voice and singer. Listeners draw conclusions about the singer's physiology and internal state based on the information provided (about both the singer and the context) in these affectations.

The listeners' ability to identify affectations is guided by foreknowledge of their likelihood (on recordings and in natural listening environments). Because listeners have previously heard many types of voices in many circumstances, they instinctively know that voices from moving bodies waver. Whispers are audible in close proximity. Present voices have energy. But not all signs are so obvious. Probability aids disambiguation. An isolated laugh may be confused for a cry. However, laughter in a dire situation is interpreted as sign of emotional instability. Since instances of instable states are rare, listeners presented with a serious situation are typically biased toward hearing a cry, not a laugh. Unfortunately, generalizations based on real-world experience do not always provide sufficient guidance for appraising recordings, which is why producers step in to move communication along.

Through music production, illusions of acoustic and sonic contexts are always artificially—even if minimally—constructed. In pop music, rarely is technology applied minimally. Amplification, signal processing and mixing makes it possible to construct not only not realistic but physically impossible, even surreal sonic scenarios: a whisper whose every word is audible above screeching electric guitars or a large chorus of angelic oohs and ahhs that steadily moves closer and closer to the listener without an audible footstep or the chafe of clothes. Such configurations defy all experiences of sound in the natural world. Yet on good recordings these scenarios make sense, the juxtapositions seem believable if not probable. Confounding predictions about listeners' appraisals, producing must create believable if unrealistic affectations. No matter how implausible or improbable, something in the sound must keep listeners connected.

Vocal affections on recordings are,

- Controlled with deliberate intention using various vocal and production techniques.
- Appraised by listeners who interpret these affectations in the context of the recording, often accepting affectations that would be/seem impossible in the natural world.
- From these appraisals they make inferences about the vocalist and the song.

3.2.8   The Producer

Producers actively *engage* listeners in making particular appraisals by offering clues, but listeners do not automatically search for the signs. Listeners have to be enticed. Shows of musical and technical prowess are interesting insofar as they deliver something meaningful, but there are other production techniques calculated to lure listeners, for example "hooks", motifs that ensnare. A hook may be part of the musical structure (such as a melodic riff), a surface feature in a performance (such as an wide interval leap) or even a non-musical attribute or non-linguistic vocalization set to music (such as a gasp, giggle or processed effect). As jargon, hooks are associated with popular music production, but they show up on all sorts of recordings, in all sorts of musical genres, for example, as unusual surround sound effects. Sometimes they are calculated, composed or accentuated; but often hooks just happen, sometimes as mistakes, at which point the producer needs to know when to not interfere.

Spontaneous exclamations, for example James Brown's immediately recognizable and pithy *Good God!*, though not intended to be hook-y, illustrate just how much information a catchy phrase can convey. Performed Brown's signature utterance usually resembles [gju:d] [gɒ/d/] (appearing on, for example, (Brown 1969, 1976). Present are affectations, non-linguistic indicators of effort, for some listeners, specific types of effort are inferred. Listeners can detect if the singer is in motion or still, his age and level of fitness. The phrase connotes "god" and all that this implies culturally, and also, a view on god signifying of a relationship between the singer and his god. As performed, the features of the words, the accents, rhythm, timbre and articulation of the utterance all very concisely provide vital clues that help listeners identify the vocalist, his state (physical, emotional, motivational and social) and interpret the song/music. Discrepancies between recorded and live performances provide another source of information. Each instance has its implications. (i.e. what does *Good God!* mean this time?) This tiny, tiny phase—consisting of two single syllable words—is packed full of information that can hook listeners' interests. Extrapolating from this, it becomes apparent the level of detail at which producers operate as well as the level of detail and nuance communicated through musical expression (and what modeling the processes of creating it entails... although much has already be accomplished). The Brown utterance is so effective at affecting it is not just memorable it is emblematic of the singer and of the funk genre.

A producer may or may not be able to fabricate performance conditions under which such idiosyncratic patterns emerge, but when they do appear, he must give them attention. Producers identify qualities that are or may be evocative in performances, and in places where such qualities are completely lacking add them by heightening or sharpening subdued emotional or otherwise meaningful displays. Hitting a pitch right in tune in and of itself does not seem like a remarkable feat, unless it is contextualized as challenging. Signal processing, effects, editing and mixing all serve to create, exaggerate or highlight salient aspects of performances and create conditions for affectations to stand out.

During the recording of David Bowie's *Heroes* (Bowie 1977), producer Tony Visconti positioned three microphones in front of Bowie at successively greater distances. The further away from the microphone a singer stands, the lower the amplitude of the singer's voice in relation to the room reverberation. The singer (singer's voice) is more distant and sounds so on the recording. For *Heroes,* Visconti controlled when each of the microphones would pick up Bowie's voice with a gate, a device that is triggered by amplitude. When Bowie sang at pre-meditated (even if only loosely) levels of loudness, one or more gates opened, and the signal at the microphone(s) was recorded. When Bowie's performance reached its most impassioned moments, his voice, drenched in layers of reverb, sounded unnaturally distant and this created a hook. As Moorefield describes it, "what the listener is left with is a strange, otherworldly quality to the vocal." (Moorefield 2005, p. 52). This affection resonates with the lyric's theme of alienation. Bowie's lyrics; the inherent idiosyncratic characteristics of Bowie's voice and vocal timbre; his performance at the microphone, the affectations he created by adjusting his voice and by using microphone technique, his pronunciation, articulation and intonation of specific words and variations in rhythmic and dynamic (and potentially melodic) structure; and the affectations introduced by Visconti's microphone technique, not only work in parallel to provide a coherent set of redundant indicators and clues but they fuse. In the final recording, the playing of the acoustic vocal instrument and the playing of the technology are so interrelated, it would be difficult to fully separate the mediated and unmediated aspects of the performance.

The producer's most valuable studio effect is affectation. The producer

- Uses affection to engage or to heighten the clarity of exposition or emotional display.
- Attends to non-linguistic, non-lyrical and non-musical aspects of the performance as much if not more than the linguistic and the musical ones.
- Uses qualities in the natural voice and/or recording and signal processing techniques to create engaging sounds. Often, these modifications enhance the perceptions of a narrative in the lyrics.

3.2.9   The Singer

The singer by performing invites subjective interpretation. Cumming's asserts,

> To a musician, it is hardly news that subjectivities may be attributed to the music itself: tones of voice, with their emotional connotations, appearing in sound; affective states, suggested by gestural action, heard in the shaping of a melodic segments; aspects of willed direction found in the impetus of tonal harmony. The musician's work is to master these potentialities on a given instrument and to work with them in accordance with the requirements of style, drawing out the possibilities of a composed musical moment by making his or her own choices of sound, emphasis, and tempo. (Cumming 2000, p. 9)

The sum of these choices results in an artistic interpretation. Through interpretation, singers give a particular voice to the song. Cone has likened this interpretive voice to a "vocal persona". As though performing an operatic role, the vocalist sings the part of the protagonist. (Cone 1974, p. 21) Of his song *I Guess I Should Go to Sleep* (White 2011), singer/songwriter Jack White says, "I have to find a reason why this person should go to sleep. What's wrong? You know, is this a positive thing or a negative thing? And try to look at it from different angles and what this character could possibly go through to get to the point where he would say this out loud." (National Public Radio 2012a) In drawing such conclusions, he says, "I get to become a director. I get to become a playwright. I get to become a producer, and an actor. All at the same time." (National Public Radio 2012b) White's vocal performance functions to relay those reasons. The choice of vocal affectations is motivated by the choice of reasons.

The Body of the Persona According to Frith, the vocal persona is of central importance in pop music (Frith 1998). "Central to the pleasure of pop is pleasure in a voice, sound as body, sound as person." (Frith 1998, p. 210). To perceive a body, information about that body must be present (in the recording). The sounding, singing body has gender, age, fitness and energy. It has a distinct physiological and emotional state, and its behaviors are understood as responses to its context (or in more biological terms, its environment). The producer influences all these attributes.

> When Sam Phillips wanted to audition Elvis Presley, he did not simply want to hear him sing. He left that to guitarist Scottie Moore… Phillips invited Elvis to his studio to 'see what he sound[ed] like coming back off the tape'. Presley's recorded voice was to be the central character in a dramatic production, and what concerned Phillips was the transmutation of the young singer's presence into an electronic persona. (Albin 2001, p. 13)

The recording medium provides both the means to construct this body and well as the shell within which to store it. The form is (re)animated through reproduction technology. "Machines are projections," suggests Eisenberg. "We create a thing in our own image, then are shocked that we resemble it." (Eisenberg 2005, p. 192). It is shocking or at least affecting because the projection calls to and engages the listener using the exact same channels of communication used for sustaining (real-time) social interactivity; and unlike television or film, which shows the listener a sound source in the form of two dimensional bodies on a screen in another place, the recording places full-sized or larger voices in the listener's space. To the ear, the distinction between a recorded and reproduced high fidelity sound source and an organic sound source is a fine one;[5] and auditory perception has adapted to respond to indicators and cues on these channels precisely for the evolutionary advantage that complex social interaction confers. As a result (of biology), some indicators and cues attract attention, and it would be quite literally unnatural not to respond to them.

But, to whom do listeners respond? The recorded form embodies both singer and persona. Embedded in the projection, coupled to any intended, contrived affectation in the timbre of the voice, is information about the *real* singer's—not the persona's— actual (not performed) internal physical state, emotional state, intentions and personal musical style. Interestingly, in popular music, the distinction between vocalists' true personas and the personas characterized by the narratives they sing is frequently ambiguous. Moreover, if the singer (not the persona) is in fact not real, listeners are likely to detect something is amiss. This is part of the appeal and the technical challenge of synthetic pop singers such as Hatsune Miko (Crypton Future Media 2007), who demonstrates her producers' flair for toying with listeners' expectations about authenticity.

The states and intentions of the persona and the singer must be believably coupled, and so not every singer is suited to every song. Not every physical body is

---

[5] Viewers of painting, cinema or sculpture, for example, perceive palpable differences between the artistic representation and real world objects. A photograph is flat. It does not exist in space as real objects exist. But a recording is not flat. Reproduced recorded sounds project into the real world and move around real spaces. Certainly, there are spatial cues, sonic details and issues of fidelity that factor into listeners' appraisals. Many recordings intentionally create artificial sounding or archetypical acoustic spaces, and invite listeners in to virtual environments. But sound quality and fidelity notwithstanding, to a listener with their eyes closed, psychophysically speaking, there is little difference between a keyboard player playing through an amplifier in the corner and the same performance playing back off of a (high-fidelity) recording coming out speakers from the same location.

capable of producing the affectations of every persona. Not every action or response is convincing coming from every body type. That is not to say a singer cannot carry a persona in the opposite gender, of another age or with a distinctly different social perspective; but the appropriate affectations must be present and integrated believably. There are countless examples where this is successfully achieved. A case interesting for the way the persona and vocalist are set against each other is Shirley Temple's *Goodnight My Love* (1936). The child singer adopts the perspective of the mother/parent with a lilting compassion that is underscored by a very adult steadfastness. The maturity of the mimicry coming out of such a young mouth is surprising. Irony makes these types of interpretations delightful; but to appreciate that irony, the listener must be able to detect both the singer's real identity and physical form while simultaneously recognizing the persona from the contrived affectations. Ultimately, the listener is not interested in the child's ability to mimic an adult, but whether the child genuinely feels what the adult feels. The listening audience decides what is believable. The young English Mick Jagger proved to audiences that he could sing like a wizened weathered American bluesman. He was convincing because he adopted affectations that both he and the great bluesmen could both render convincingly.

Affectations in the vocal performance

- Offer listeners clues that convey information about the meaning of the song, the context of the performance and the vocal persona.
- Convey the internal state and intentions of the narrative voice as well as those of the singer.
- Whether indicative of a persona or the singer are recognized by the audience and deemed believable or not.

Listeners know recordings are contrived, and that knowledge comes to bear on appraisals; but nevertheless produced vocal recordings affect us deeply, immediately and instinctively.

3.3   Instinct and Non-Linguistic Communication

Non-linguistic vocal communication is incredibly flexible and has incredible utility. Not only is subtle, effective and affective communication possible with pre-verbal humans, it is ultimately necessary for nurturing them into linguistically adept, socially integrated humans capable of sustaining interaction. We communicate with infants to teach them how to behave and communicate. "Communication is first and foremost a social event, designed to influence the behavior of listeners." (Seyfarth 2003, p. 147) Sound is one vehicle for delivering messages, but the animal kingdom is rife with species that display behaviors, secrete chemicals and/or vocalize expressly for the purpose of influencing the behaviors of others. Socialization and survival are intricately enmeshed for many, many species.

Biologists refer to this type of communication as "signaling". The human species uses the voice for many kinds of social signaling. Mating calls, bee dances, territorial displays, these are signals— that get other animals to react and exhibit behaviors. Maynard-Smith and Harper distinguish between signaling and coercion. Signals beget active responses. If pushed, an animal is coerced. If a roar causes it to (take action to) retreat, it is a signal. The retreat response is linked to the departed party's perceptual abilities, which are tuned to detect the meaning in the roar. Such response behaviors are the product of evolution. Maynard-Smith and Harper distinguish signals from cues as well. Signals are produced intermittently as circumstances demand. Cues, though also features that influence "future action", once produced persist without additional effort or cost (in terms of energy, health, etc.). (Maynard-Smith 2003, pp. 3–5) For example, there are in humans vocal cues such as fundamental frequency indicative of the gender of a speaker/singer; and also signals, for example, in timbre and prosody that indicate stress.

Signals may announce intentions or precede actions. Signals may be aimed at conspecifics or across species. (Krebs 1984, p. 380) Animals have repertoires of signals that enable them to procreate, to threaten, to alert and to bond. Vocal signals are identified by characteristic patterns (possibly discrete cues, possibly relationships among dimensions) that may be embedded within already meaningful calls. For example, a birdcall may signal that the singer is looking for a mate. Conspecifics understand what the melody means, or at the very least, respond to the information imparted through it. Potential mates are attracted to or dissuaded by what they hear, and react accordingly. Some signalers are more successful at attracting mates. Within any particular signaler's call, in the vocal timbre, energy, loudness, etc., are telling sonic patterns which signal the signaler's state of health, intentions, social status and some would argue (in some species), emotional state. Goodall contends, "The production of sound in the absence of the appropriate emotional state seems to be an almost impossible task for a chimpanzee" (Goodall via (Tomasello 1997, 2008 p. 258)); which demonstrates the weight of emotion in governing behavior.

Krebs and Dawkins interpret the purpose of signaling cynically (their description), and suggest that signaling is the means by which animals exploit other animals. That is signalers *manipulate* "victims". Other animals, such as potential mates, and inanimate objects, such as sticks, are equally exploitable as tools for survival. (Krebs 1984, p. 283). However, a manipulator's relationship with an animate victim, capable of autonomous action (and retaliation), is significantly more complicated than a relationship with a stick. In the biological sense, the exploitation of victims is generally associated with behaviors involving muscle power such as backing out of territory or bearing young. In some situations, such as mating, the victim (and species as a whole) stands to benefit from being manipulated.

In a study on house cats, McComb et al. investigated (McComb 2009) the use of purring to solicit food from their caretakers. They recorded 10 cats' purrs in food solicitation and non-solicitation interactions, and asked 50 (human) subjects, not all of whom had prior cat care experience, to rate the purrs for urgency and pleasantness. Subjects rated solicitation purrs as more urgent and less pleasant, both when they compared the purrs of one cat and when they compared purrs among cats. Purrs are low in frequency, usually with a fundamental around 27 Hz (toward the low end of the human hearing range, estimated to be 20 Hz to 20 kHz). In an analysis of the acoustic signals, McComb et al. found in the solicitation purrs pronounced peaks falling within a frequency band of 220–520 Hz (with a mean of 380 Hz). They re-synthesized the urgent purrs, removing these peaks, and played the modified stimuli alongside the original stimuli for subjects. The re-synthesized material was rated lower in urgency. The researchers, offering a potential reason for the signal's efficacy, note that the cries of human infants usually have fundamental frequencies in the 300–600 Hz range. It is also worth observing, if this is manipulation, most cat owners reap benefits from being manipulated, at the very least because they enjoy the experience.

Seyfarth and Cheney stress that there is a clear distinction between signaling and informing. When an animal calls, it is not its intention to give away its location to a predator, although this happens. (Seyfarth 2003) Signals are not intentional in the same way that an expressive musical gesture is intentional, but singers do signal and may send unintended signals and unintended messages. As compared to communication in the wild, in the recording studio it is possible to give signals and messages special attention, for example, by allocating the responsibility to a producer who is sensitized to listens for them. Indicators that appear encrypted in musical gestures may be intentionally left in with the expectation that certain listeners will derive meaning from them. The "gangsta" rapper intends to use language and symbolic, expressive gestures that (wannabe) gangsta listeners are likely to recognize and understand. However, gangstas and non-gangstas alike can perceive the indicators of aggression, dominance and territoriality in the music, and from inevitable signals in the voice, make inferences about the rapper's level of energy, physical fitness, social standing, age, sincerity, etc., and different listeners respond differently to this information. For this reason, the producers' abilities to detect and infer what "victims" will hear and their ability to intercede as necessary all serve to enhance communication between signalers and victims.

Natural selection favors adaptations that improve the production and perception of acoustic patterns related to signaling. Seyfarth and Cheney have identified an "Audience Effect" in animal vocalizations. Signals appear when there are listeners with whom to communicate. (Seyfarth 2003, p. 147 and 150) Adaptations in cat behavior enhance communications with humans, which improve cats' chances of survival. Signals that communicate best and most effectively elect beneficial responses perpetuate. In order to respond to signals, the manipulated animal must be able to detect the signal and make meaning of it. Krebs and Dawkins describe the detection of signalers' physiological and/or emotional state and the prediction of signalers' actions as "mind-reading". It is an aptitude that offers would-be victims a distinct evolutionary advantage in terms of socializing, hunting and self-protection. (Krebs 1984, p. 381) Humans who read their cat's behavior better socialize better with their cats.

3.3.1 The Environment and Objects in the Environment

Marler, Evans and Hauser maintain that some animal signals communicate information that relates not only to themselves, but also the environment. There is a "relationship between the signal produced, the environmental context and the

caller's motivational state." (Marler 1992, p. 66). This suggests that this information is available for utilization by mind readers in appraisals of the signals and the
signaler's behavior. These different aspects of a signal are distinguishable. Referential and affective qualities are distinct, but references can also be indicative
of emotions (Seyfarth 2003). Regarding human communication, Clark elaborates,
"think of demonstrating, indicating, and describing-as, not as types of signals, but
as methods of signaling that combine in various ways." (Clark 1996, p. 161) "Every
signal, every bit of language use, occurs at a particular place and time. They need
to be anchored to that place and time." (Clark 1996, p. 165) Communication among
adult humans fluidly combines signals indicative of internal state, motivations and
the impact of the environment on the signaler with external references to the context and objects or states in the environment. (Internal and external references are
less obvious in animal communication and in interactions with human infants.)
Signals are not language. They do not usurp linguistic communication in human
social activity. It is rather that "language use could not proceed without signals".
(Clark 1996, p. 155)

Production can be viewed analogously. Every musical pattern and lyric is coupled to a performance and every performance with a recorded context. Producing
constructs an acoustic and social environment for that context and orchestrates and
arranges it. The recording therefore provides its own reasons why a singer is signaling or indicating certain information. The varied constituent components inform
listeners as they make appraisals. If that information is contradictory or does not
lend itself to coherent interpretation, then listeners are prone to disregard the communication. Listening uses precious energy, which in terms of (biological) survival
is a cost. Why listen to noise? Noise confuses, and is not reliable as means to elicit
specific responses.

3.3.2 Cooperation

Like other species, humans are constantly involved in predicting and reacting to the
signals transmitted by others, assessing threats and detecting opportunities. "Natural selection will tend to favor animals that become sensitive to available tell-tale
clues, however discrete and subtle they may be" (Krebs 1984, p. 387) because survival depends on the ability to socialize successfully, to choose whom to trust and
to avoid and to coordinate mutually beneficial activities. In the social context of
the recording studio, as in many business environments, the producer consciously
"reads the minds" (in the Krebs and Dawkins sense) of the clients he produces,
detecting changes in mood, level of confidence, energy level and attention that may
impact a performance. These signals, crucial for facilitating and coordinating the
session, are not the utterances the clients speak or the music, but encoded in the utterances and performances. Cooperative behavior of any sort in any species requires
signaling to synchronize and regulate behaviors. Signals indicate who is ready to act
and what types of actions are likely; and also how a group is organized, who are the
alpha leaders and who are the beta followers. Signals offer clues into a signaler's

expectations and needs, and also, probable success in any given action. In the recording studio, if these signals go undetected or are ignored by collaborators, social interaction and ultimately sessions fall apart.

- Animals signal to influence the behaviors of other animals. Krebs and Dawkins have described this as a form of "manipulation".
- Animals adept at reading signals, "mind-reading," have an evolutionary advantage. They are better socializers, hunters and self-protectors.
- Signaling is essential to all cooperative behavior, including making recordings.

Cooperation involves, if not trust, belief.

3.3.3 Faking It

> The most potent way to tell a story is to first have lived it. And if you lived it, then you're speaking the truth. And when you speak the truth a listener will feel that it is true. And therefore, your story will be more contagious. And I think that's what listeners look for in songs. They want to feel that the writer, the singer, has lived it; and then they can sort of experience the same feeling by entering the song themselves. —producer Daniel Lanois (Frenette 2007)

Mind readers will not respond if the manipulator's signal seems implausible. There is little value evolutionarily speaking of erroneous information, other than to identify the signaler as not credible and untrustworthy. Nevertheless, humans and other species do lie. Reddy observes, "Even [human] toddlers can lie, flexibly, non-formulaically, and for interesting 'psychological' reasons. Such deceptions have their roots before speech, where, from before the end of the first year, infants are concealing and faking and distracting and pretending in their non-verbal interactions with other people." (Reddy 2008, p. 215) Animals (humans included) learn from experience, and so learning how to lie comes from social interaction. As we learn what is acceptable behavior, we learn the limits of acceptable, believable deception. (Reddy 2008, p. 222) We learn that lies work sometimes and that lying has benefits. Unfortunately, there is a lot of bluffing in record production. There is the desire to reap the rewards of being someone else, someone more talented or with different life experiences. There is peer pressure, and the expectations of those paying the bill for studio time. Producers scan for deceptive information before the audience passes final judgment; but sometimes some gets through.

If animals (humans and others) can deceive and know which signals produce the desired result, if growling a particular way offers an advantage, what is to prevent or discourage dishonest signaling? Grafen poses a "handicap principle" (Grafen 1990; Fitch 2002, p. 66), the idea that signaling bears a cost for the signaler, typically in terms of expenditure in energy, that discourages superfluous signaling. For humans, risking credibility is dangerous and loss of it bears social costs that leave deceiver vulnerable. Clutton-Brock and Albon observe that many honest signals are dependent upon the animal's physical condition and stamina and thus acoustically speaking cannot be faked. (Clutton-Brock 1979) In some animal calls there are correlations between, for example, the animal's size and features in the sounds they make. Low frequencies come from large-sized objects. Specific sonic features map directly to certain meanings. (Fitch 2002, p. 70) Mind readers (human and otherwise) instinctively seek congruence in what they hear since perceiving involves the inferring of causal explanations. Stereotypically, opera singers are depicted as large because slender petite opera singers with low, powerful voices, no matter how beautifully they sing, no matter how honest the musical gestures, no matter how real the performance, produce cognitively dissonant signals.

3.3.4   Convincing

On the other hand, Picasso claimed that art is lies that tell the truth. In some circumstances, both humans and animals are successful in their attempts to provide misinformation, even to conspecifics within a tightly knit social group. Animals sometimes fail to signal when signaling is required (Fitch 2002, p. 71), for example, when food is available (and might be shared among the group). For an animal's lie to be convincing, according to Fitch and Hauser, it must satisfy three conditions: a given call (within a given species) must be tightly correlated to a specific, common context (i.e., the presence of a specific predator); the signaler must expect the victim/mind-readers to respond to the dishonest signal in fairly predictable ways; and there must exist within the species the ability to manipulate other members of the group by using common calls in new situations. (Fitch 2002, pp. 107–109) Wiley stresses the relative rarity of deceitful signals. Communication requires rules to govern the encoding and decoding of information. A successful deception violates established rules (Wiley 1983, p. 188), and must not happen often. Exceptions are rarely available to be detectable, keeping the patterns of deceit unfamiliar.

Deceptions nevertheless may have distinguishing traits. In an early study on the impact of motivation on language usage, Osgood and Walker examined genuine and fabricated suicide letters, looking for indicators that would distinguish them. (Osgood 1959) The sample consisted of 33 pairs of real and "control" letters penned by both men and women. Measures for comparison were developed in a pre-study that compared (genuine) suicide notes to standard letters. The differences they identified could be explained by circumstances of high motivation and stress. (Motivation and internal state are signaled.) Real suicide notes were repetitious and showed less lexical diversity; they had fewer adjectives and adverbs; they used shorter words; they contained more demands and commands, referred to as "mands"; and they used "extreme", "allness" language. Osgood and Walker borrow the "mand" terminology from Skinner. A mand "is an utterance which (a) expresses a need of the speaker and which (b) requires some reaction from another person for its satisfaction. It is usually expressed in the form of an imperative" (Osgood 1959, p. 62). "Allness" refers to "polarizing" language such as "always", "never", "forever", etc. It should be mentioned that modern literature brings into question the accuracy of some of these measures for validating suicidal intentions in clinical and legal settings. Regardless, suicide notes that are more affective at delivering their mands by whatever means, are more successful at electing reactions from readers, be it in the form of interventions, sadness or guilt. Successful suicide notes manipulate.

Of greater interest here, however, is the basic approach to methodology. Measures were identified for empirical tests; and also these were measures expected to correlate to motivational state. Such approaches point the way to models of communicative behavior that incorporate the influence of context and motivation. Also, it cannot be missed that several of the qualitative measures Osgood and Walker identify (as indicators of motivational state, not suicidal intention) lend themselves to comparisons with qualities found in music, lyrics, performances and productions, measures such as redundancy; structural disturbances; average length of independent segments and qualification of verb phrases (for example through emphasis). And, pertaining to lyrics were the "mands", for example, in the song *Hang with Me* (Robyn 2010); and also, "allness" as in *Every Breath You Take* (The Police 1983) or *All I Want* (LCD Sound System 2010) These particular attributes, according to Osgood and Walker, are indicators of high motivation states, and thus (whatever the motivation) it makes them good fodder for pop songs.

Osgood and Walker's analysis aimed to identify which suicidal characteristics could be faked and which could not. They found that notes generated "on-demand" incorporated standard themes such as "taking a 'way out'... asking forgiveness... saying 'farewell'...moral and religious implications". (Osgood 1959, pp. 65–66). Faked notes had fewer mands, and ambivalent constructs such as "maybe", "appears" or "should"; and more adjectives and adverbs. The authors hold that although the distinctions are small, they are discernable. The statistical analysis revealed several quantifiable differences when the authors and eight graduate students (blindly) categorized the letters in the sample as either genuine or fake. The authors' categorizations were correct 31 out of 33 times and 26 out of 33 times. The graduate students, however, on average were correct only 16.5 out of 33 times, illustrating that familiarity with the materials impacts appraisals.

A similar effect is found in music listening. Not only will listeners' formal knowledge of music impact how they interpret what they hear, but familiarity with a genre will also influence attention and what in a performance is considered significant. Lerdahl and Jackendoff suggest, "Given that a listener familiar with a musical idiom is capable of understanding novel pieces of music within that idiom, we can characterize the ability to achieve such understanding in terms of a set of principles, or a 'musical grammar', which associates strings of auditory events with musical structures." (Jackendoff 2006, p. 34) Capacities for music understanding, in some ways mirror capacities for language. Listeners know the rules and recognize them. Rules can be modeled and turned into procedures.

However, grammatical representations are somewhat self-contained. The rules for decoding the patterns are embedded within the music itself. Kramer calls this a modernist view. He claims, "Modernist forms of musical understanding ascribe a unique self-referentiality of music that renders it largely opaque from 'extra-musical' standpoints. Music must somehow be understood from the inside out." (Kramer 1995, p. 13) Alternatively, Kramer emphasizes the importance of cultural experience and knowledge, arguing that music is heard through a "broader field of rhetorical, expressive and discursive behaviors." (Kramer 1995, p. 31) From this perspective, listeners' experiences of voices in social contexts and interaction come to bear on appraisals of recorded performances. Dibben (2001) proposes that while people listen to the sounds within the music itself, they understand that these sounds simultaneously refer to objects outside the music, objects with cultural significance. Recorded vocal performances, recordings in general, intentionally reference recognizable cultural artifacts. A singer may imitate another singer or performance style of another era (e.g., Amy Winehouse). Recorded voices are processed to sound like old radio broadcasts or robots; etc. Producing introduces these references and/or makes them overt, utilizing and enhancing their social significance... to affect.

3.3.5   Social Interactions

Interacting socially is significant and affecting because it is deeply ingrained. It is an evolutionary birthright. Trevarthen asserts that humans begin life as social, albeit non-linguistic, creatures. "A human being is born capable of seeking and playing with others' attentions and feelings in a rich variety of provocative, humorous and teasing ways." (Trevarthen 2011, p. 121; Reddy 2008) Produced recordings function similarly. Off the record, conditioning and convention dictate that there are appropriate ways to interact socially. In conversation, it is indecorous to weary listeners with unnecessary repetition, hyperbole or unwarranted emphasis. Such things provide no useful information. These social maxims have counterparts in produced vocal performances. As they are played, recordings continually engage listeners' interest, advancing communication with new information, new patterns and signals. The information reinforces the relevance of the message (for the titular interlocutor). A mismatch between the nature of the information and the amount of effort required for making appraisals of it leads to communication failures. Since listening expends energy, undue cognitive demands are costly. Overly exaggerated or disingenuous performances, incessant pleas for attention or a prolonging a sense of urgency, over time divert focus from information that makes messages personal and interesting. It discourages listeners from participation, and perhaps unintentionally, adds emphasis to the song's and/or singer's self-importance. Any of these may be considered a social faux pas, and are also discouraged in production.

As Wittgenstein captured in the "language games", in verbal exchanges, language usage and its interpretation are dynamic. They vary under the influence of circumstance, just like vocal performances and appraisals of a vocal performance. Innately, humans adjust communication patterns to suit particular social situations, to minimize communication failures and reinforce social cohesion. (Labov 1973) cited in (Velleman 2007, p. 37). Producing adjusts patterns of communication to improve congruence among vocal behaviors and the musical and social context, and thereby minimizes communication failures and reinforces the social cohesion.

Affectations provide a vehicle through which the listener, recording, performer and perhaps songwriter connect, if the listener is given opportunities to do so. Although, physically and temporally detached, entities are able to form an illusory social bond, momentarily as though in a single instance of interpersonal communication or over the long term, turning the listener into a fan. Listening audiences form cohesive social groups with identities. Their members share common cultural reference points; they participate in similar cultural activities; and are guided by similar social constructs, all of which serve to make music meaningful.

From an evolutionary perspective

- Signaling is essential for all coordinated social activity.
- Not all signals communicate information about emotional state. Some pertain to physiological state, motivations and intentions.
- Signals may be internally or externally referential.
- Signals communicate information about the signaler's environment.
- Faked signals are possible.
- Vocal performance provides a vehicle for signaling.
- Producing, which enhances affectations, also enhances signals.

There are signals about the singer's state in the singer's voice. What producing does to the voice, it does to vocal signals. It is important to differentiate, as Seyfarth and Cheney do, signaling from informing (Seyfarth 2003). A vocalist's signals are not always matched or even congruous with the singer's artistic intentions. Recording and music production affords unusual opportunities to analyze, monitor, filter and shape communications between signalers and listeners (victims). By representing an audience's perspective, the producer perhaps encourages signaling. Signals in the animal kingdom are not intentional, but in the studio the producer unnaturally enhances and contrives situations under which signals appear, thereby bringing them under the influence of intentionally decision making.

Listening to recordings is like a social activity in that

- They engage listeners and sustain attention with indicators and cues used to initiate, sustain and coordinate social interaction. Some of these indicators foster social bonds.
- Listeners bring to their appraisals an understanding that sounds within a recording refer objects with social significance outside the recording.
- Listening behaviors mark groups bound together by similar social constructs.

The producer through his choices and leadership aims to communicate the identity of the artist, what the song is about, the identity of the vocal persona, the persona's circumstance and the artist's relationship to the song. Sonic features in the voice give rise to these representations. Whether of biological or cultural origin, many affectations in sound recordings have social significance. That abilities to render these qualities vary explains why some productions and some producers are more affecting.

3.3.6 Manipulating Listeners with Recordings

Birds of the same species sing the same song, but all performances are not equally alluring. Some singers' songs have attributes that listeners find more desirable.

Vocalists, whether avian or human, may or may not be aware of the cues and signals being displayed, or be able to exercise control over them. Enter the producer. In a cynical read on record production, the producer recognizes these engaging attributes, and by exaggerating them and/or making them more perceptible (through musical performance and by technical means) intends to bring about changes of mental state and possibly behavior in listeners. The producer may be explicitly aware of this manipulation or merely act out of instinct— because the affect has proven advantageous.

As manipulator, as intermediary between the performer and audience, the producer is uniquely sensitive to communication failures. Messages must get through to have affect, and what gets through in live performances differs from recorded ones. "What performers notice [in recordings] are not errors but aspects of style or interpretation. What may have felt right in the heat of the performance may in retrospect sound overdone and contrived or, at the other extreme, flat and lifeless." (Katz 2010, p. 33) The singer and the producer add appropriate emphasis to the features obscured and de-emphasize those that as a result of being over-amplified by the recording medium confuse the listeners' appraisals. Within these limits, which are not fixed but relative to the constructed recorded context, affectations that communicate important information are detectable, recognizable and believable.

Messages elicit greater responses when amplified. Producing engages but moreover heightens the listeners' response. In a classic animal study on perception and behavior, Tinbergen observed seagull chicks pecking at their mother's beak to solicit food. Seagull beaks have a red dot toward the end. Tinbergen created cardboard gull head and beak models. On some he painted dots, in varied colors, on the end of the beak. Some he left without dots. He then presented the artificial stimulus to the chicks to see if they would peck; which they did. Some models elicited a more vigorous response. The dot pattern triggers an "innate releasing mechanism." (Tinbergen 1948) via (Cate et al. 2009, p. 795). Initially, Tinbergen reported a preference for red dots, but this finding has subsequently been questioned. He and others have since suggested that the chicks respond to the contrast pattern. (Cate et al. 2009). In a later study, Tinbergen presented the chicks with a beak model containing three dots, and this elicited an even more vigorous pecking response. (Tinbergen 1954) via (Ramachandran and Hirstein 1999). Ramachandran calls this a "super stimulus" and suggests that the brains of gulls and other animals including humans may have a "more is better" rule that helps to regulate response behaviors. Surveying the prevalence of exaggeration in art, Ramachandran speculates that this may provide partial explanation for our preferences in artist expression. Artistic representations, he suggest, are super stimuli that excite heightened neural activity. (Ramachandran 1999), also see (Dissanayake 1995, 1988) for an additional discussion about the exaggeration in play and the attribution of specialness in relation to art)

Human vocalists exaggerate features during performance to exaggerate the saliency of the gestures and the affects on listeners. Recording technology and producing with recording technology afford additional tools for amplification, but also, the most basic properties of the recording medium itself encourages heightened, exaggerated responses. In pop music, like three red dots, the voice which is indicative of a subject, placed front and center, demands listeners' attention. Frith sees the recorded voice used "as something artificial, posed, its sound determined by the music. And technology, electrical recording, has exaggerated this effect by making the vocal performance more intimate, more self-revealing, and more (technologically) determined." (Frith 1998, p. 210) By granting exceptional access to the performer, recording brings the listener into a context of familiar, intimate exchange, and thereby ups the antes for listener response.

Listeners display behaviors in response. To participate in the activity, the listeners must attend for sustained periods of time and part with resources for the privilege (i.e., energy, time, and in due course, money). Either they do so willingly because experience has shown that there is a benefit to this response or they do so as a result of conditioning or adaption. In this way, production evokes or provokes reactions and changes listeners' behavior, sufficiently at least, to sustain engagement with the recording.

Being manipulated has potential benefits. There is an overwhelming amount of anecdotal evidence to suggest that music and music recordings affect us deeply. The benefit may be to "sort of experience the same feeling by entering the song themselves" (Lanois via (Frenette)). Kivy suggests there is an important distinction between perception and induction "Two alternatives seem available: the view that music is sad in virtue of arousing sadness in listeners (what [Kivy calls] musical emotivism), or the view that music is sad in virtue of possessing sadness as a quality that we can hear in the music, not that we feel in ourselves (what [Kivy calls] musical cognitivism)." (Kivy 1989, p. 154). Current empirical research seeks to ascertain if listening brings about measurable changes in the listeners' mental and/ or physical state. (For an introduction to music and physiology see (Schneck 2005) and music and emotion (Juslin 2010)). Whether the effect is direct or indirect; by attracting attention with meaningful sounds; by using signals that humans use to navigate social activities and with other recognizable cues; the features of performances are being intentionally configured and shaped through production to impart information that is expected to have a manipulative affect on listeners.

From a socio-biological perspective, the goals of producing a vocal recording may be summarized as follows:

- To clearly represent the internal state and the intentions of both the narrative voice and the singer on the recording.
- To create messages which are coherent and easy for many listeners to recognize, often by using exaggeration.
- To use signals, cues, indicators and signs with meanings commonly understood among members of the same group.
- To draw attention and provoke a response.

3.4 Features and Their Functions

Music production is predicated on methods (and formulae). New tools and techniques appear periodically to facilitate those methods, enhancing the advantages they impart, their efficiency and their sustainability. Production's evolution depends on successfully emulating established processes and adapting them for new contexts. Affects and affectations are observed, measured, modeled and synthesized by producers, and the same may be undertaken by others who would emulate the producer's processes. The more insightful and meaningful our analyses of productions, producing and producers, the more the accessible the process of producing becomes and the greater the potential for acting on it technically, musically and artfully. Future research into producing is essential for realizing this potential.

3.4.1 Identifying Features in Producing, Producers and Productions

Though there is no single set of parameters relevant to every production and every producer, within any producer's body of work patterns and typical configurations of those patterns surface. What combination of patterns constitutes a producers "signature"? How long is a signature pattern in recorded music? How many times does it appear in a recording? Does it appear within a single parameter or dimension or is it a composite of patterns that span parameters, and which parameters are likely to bear signatures? And which aspects of musical performances carry those influences: unusual juxtapositions of emotions or other signals; digital effects processing; rhythm; dynamics; acoustics (real or artificially rendered); etc.? Also, in what contexts do these patterns appear?

But for all these patterns that speak to consistency, producing makes the most of discrepancies. Recordings share similarities while remain distinctly varied. Nevertheless, variances in verbal/linguistic communications, signals, cues, musical gestures, and all the other musical characteristics that factor into a performance, though complex and interrelated are distinguishable and can be compared.

Analysis and Synthesis To emulate producing methods (intuitively or by implementing the technological means to do so) requires analysis and synthesizes. The evolutionary and socio-biological perspectives outlined above bring forward parameters, affectations and sonic features that are significant and meaningful to a broad spectrum of listeners, across genres, stylistic conventions and outside the contexts of specific songs. These observations do not replace earlier models of music making but expand on them by providing a frame through which

- Affectations, cues and signals embedded in produced performances may be identified and analyzed, building on existing research in non-linguistic, vocal communication.
- Sensitivities to types of signals, cues, affectations and information delivered through specific channels/parameters/dimensions may be evaluated.
- The affect of indicators may be observed as functioning relative to specific contexts.

Producing may be described as a process of

- Detecting pertinent information that is poorly communicated or obscured by inarticulate affectations or noisy, faint signals.
- Choosing parameter configurations that lend themselves to interesting and convincing affectations.
- Choosing affectations and signals that are likely to evoke responses.

Emulations of particular producers' styles may incorporate

- Affectations and indicators that, as evidenced by the prevalence of these features in a corpus of the producers' work, the producer is likely to emphasize.
- Typical configurations of parameters representing how the producer realizes particular affectations and signals in different types of contexts.

These indicators, affectations, signals and configurations may use features not specified in music theories. It is also possible to evaluate recorded productions in terms of how they engage listeners and draw attention, sustain interaction, facilitate bonding and social cohesion, and how through the non-linguistic aspects of vocal performance these connections are accomplished.

The context surrounding the use of an affectation under study may also be deconstructed, making it is possible to describe

- The circumstances under which an affectation or signals is likely to appear as well as how often it is likely to appear.
- The inferred meanings of specific affectations and signals relative specific contexts.
- The effectiveness of affectations and signals within a context.

With these baselines more meaningful comparisons may be made among productions and also producers, and vocal qualities may be compared to other instruments and sonic effects.

Whatever the choices and goals that lead to the discernable characteristic of recordings, as a matter of course, producers employ tools to render their productions. The effects of mediating technologies on what is produced deserve consideration as well, not least of all by those designing new tools. Producers are (by nature) both pragmatic and inventive in making the most of what is at their disposal. The more tool designers understand about the artistic and technical goals, the desired outcomes and the established methods or techniques of the production process, the more they can reliably predict how a given tool may be used or (intentionally) misused (to innovate), and how familiar functionality may be extended.

3.4.2 Tools and Production Methods

Certain functionality is required in the studio production environment. Needs include the means to record, edit and mix sound; to control the balance of instruments/ sound sources and position them in space (by feeding them to multiple speakers, in mono, across a stereo field, in surround sound, etc.); and to apply various forms of processing to individual and groups of sounds. Production also requires the means to generate sounds and musical patterns and control their (musical) performance. For these purposes, different technologies are employed at different stages throughout the production workflow.

Any time a wholly new type of tool is adopted or incorporated, workflows change to accommodate it. Analog studios were largely comprised of discrete specialized devices, each of which performed singular or a relatively small set of functions. Devices were chained together and audio signals were routed through them. The vestiges of an analog approach to system design are still to be found in digital production environments, mostly in the interfaces rather than at any operational level. In the digital domain, typically, a single (computational) system houses integrated components. Most digital audio workstations and audio software applications provide a suite of tools, meeting multiple varied common needs. During production, specialized applications may be added/accessed without necessitating that the audio signal leaves the system (or the digital domain). Generic workflows incorporate the following:

Recorders and (non-linear) editors such as *ProTools* created by Brooks and Gotcher (first released in 1991 (Avid 2013)), which is so ubiquitous as to be almost a de facto industry standard, though there are alternatives. *Audacity*, a freeware application created by Mazzoni and Dannenberg, (Mazzoni 2002) has made multi-track recording accessible to a great many amateurs, experimenters and academics. It has been widely utilized, though rarely in commercial production facilities. Hard disk recorders and editors usually offer integrated signal processing features. It is generally understood, however, that users will want to customize and extend systems with plug-ins.

Signal processors and audio effects are an integral part of the record production process. They are as (musical) instruments that offer unique timbres or creative possibilities. As software and plug-ins are less expensive than analog devices and effectively occupy no physical space, they have largely replaced outboard processing gear. Given their importance for shaping recorded sound, their individualized nature and the idiosyncratic way they are used, the number of processors a studio/sound engineer/producer/user expects to acquire and the sources from which they will be obtained (multiple companies, development communities, etc.) is rather open ended.

Some forms of signal processing are designed not to change the sound but instead to extract from the signal information about the sound's features. For example, there are various forms of loudness metering; and tools that enable metering at different points in the signal path and of different (sub-grouped) components of a mix, of peaks and of loudness integrated over time. Other signal-oriented attributes such as spectral energy are also monitored. And, various musical features such as pitch and tempo are routinely extracted computationally (rather than by ear). The information provided by these tools impacts production decisions and/or leads directly to modifications of the sound and/or performances.

In production, performances are constructed by editing and processing recordings of live performances, and also, by literally synthesizing them. Performers, producers and engineers, use various (non-acoustic) sound generators, such as software synthesizes (soft synths) and samplers (used primarily to appropriate pre-recorded, sometimes non-musical sounds for musical use). Controllers and control surfaces link live performances to software, triggering samples or effects for musical performances, or software to transport controls for engineering related performances. Composition tools such as sequencers and drum machines are used to construct virtual performances using synthetic instruments.

The commercial applications and plug-ins that provide these necessary tools for recording, processing and synthesizing are developed parallel to or sometimes as a direct result of academic research into audio signal analysis techniques, processing algorithms, programming languages, auditory perception, artificial intelligence, etc. Each in their ways makes sonic parameters available for creative manipulation and/ or enables music creators and technologists to model or codify musical processes. Many tools developed in academic environments, unrestricted by conventional production workflows, make it simple to merge composition, synthesis and processing tasks, and also, media types. For example, Puckette has developed several visual programming languages for authoring interactive music starting with *Patcher,* which gave rise to the commercial product *Max* and later an open-source version, *Max/MSP*. Puckette also spearheaded the development of *Pure Data* (Pd), another graphical programming environment for real-time processing of media (audio, video and graphics) (Puckette 1988, 1991, 1996). *Faust* (Orlarey 2006) by Orlarey, Gräf and Kersten is language for digital signal processing and is used to develop real-time plug-ins for, for example, Pd. Vercoe's *Csound* (Vercoe 1990) language is widely used for sound synthesis and electronic music composition. Similarly, McCartney's *Supercollider* language and programming environment (McCartney 2002) is well suited to real-time sound synthesis and algorithmic and generative composition techniques (procedures for rendering variation). Although these tools are very rarely found in commercial production environments, the creative and technical potential they expose influences the audio industry and music production.

New paradigms do not have a release date. They emerge. By combining record production and procedural composition techniques, similar to those afforded by *MAX/MSP* and *Supercollider*, commercial artists have created interactive music experiences that complement and/or accompany conventional studio recordings, for example, Björk's *Biophillia* app album (Björk 2011). The interaction models behind such musical apps owe something to earlier work in multimedia such as Peter Gabriel's *XPLORA1* interactive DVD (Gabriel 1993). Even as they are informed or inspired by music productions, these creative experiments offer all sorts of ways to engage with music interactively, procedurally. However, one sound, because of our highly developed sensitivities to it, remains troublesome to expose to interactive manipulation— that is the voice. Brian Eno, for one, sees many prospects for "voice shaping technologies"… in the very ways they challenge our expectations about what voices are supposed to sound like and manipulate our perceptions to create something new. Regarding his song *Bottomliners* (Eno 2005), he says, "I am using a machine which was designed originally to correct bad singing… If you over use it, the voice stops sounding human, and it becomes sort of like an angel's voice… because it is too perfect… This is interesting for me because it means that machine creates a new person really, a person who could never exist in real life; and that is much more interesting than making an ordinary singer sound in tune. So, very often what is happening with technology is that you are taking a machine to solve a problem; but using it to make something no one ever thought about making before." (Youtube video of Eno (Unknown 2006))

3.5   Conclusions

Voices are rich with affectations, signs, indicators, cues and signals that inform listeners. Affectations in sung performances convey information to audiences about the singer's physical, social, motivational and emotional state, about the song, and about the vocal persona's state, motivations and circumstances. Record production offers opportunities not available in live performance to contrive affectations that affect and provoke responses in listeners. Utilizing non-linguistic channels of communication, the same that are used for sustaining and coordinating social interaction, producing refines the display of affectations; and controls the redundancy of information that these affectations transmit, stacking the odds in favor of desired inferences and responses. Producing minimizes communication failures with listeners and strengthens (social) connections.

By viewing the process of production through a socio-biological lens, many measurable attributes come to the fore that provide a basis for emulating the effects of producing and the processes of producers. The next step in this work is empirical investigations, in controlled contexts, through which the perceptions of producers and listeners may be measured and the affects and efficacy of production decisions evaluated.

Software emulators for analog hardware are now as common in recording studios as console automation and many systems offer built-in presets or processing designed to assist users in optimizing the audio signal in terms of sound quality and musicality. These tools offer unprecedented possibilities for artistic and technical control over the recording process. Models of producing that lend themselves to procedural representation will offer yet more creative opportunities in music production.

Acknowledgements The author would like to acknowledge the support of the AIRS project, Advancing Interdisciplinary Research in Singing, A Major Collaborative Research Initiative of the Social Sciences and Humanities Research Council of Canada during the very earliest stages of this study; and Luleå University of Technology for support during later stages.

References

Albin Z (2001) Poetics of rock composition: multitrack recording as compositional practice. University of California Press, Ewing
Antares (November 2012) Auto-Tune 7 Product page. <http://www.antarestech.com/products/auto-tune-7.shtml >. Accessed 30 Nov 2012

Austin L, Cage J, Hiller L (1992) An interview with John Cage and Lejaren Hiller. Comput Music J 16(4):15–29

Avid Audio (15 January 2013) ProTools product page. Avid Audio. <http://www.avid.com/US/products/family/pro-tools>. Accessed 15 Jan 2013

Bachorowski J, Owren M (2003) Sounds of emotion: production and perception of affect-related vocal acoustics. Ann N Y Acad Sci 1000:244–265

Baddeley R, Hancock P, Földiák P (2008) Information Theory and the brain. Cambridge University Press, Cambridge

Balaban M, Ebcioglu K, Laske O (1992) Understanding music with AI. MIT Press, Cambridge

Björk (12 October 2011) "Biophilia". Biophilia. One Little Indian, Ltd. and Well Hart, Ltd., San Francisco

Bowie D (1977) "Heroes." Heroes. RCA, Vinyl record

Bradbury J, Vehrencamp S (2011) Principles of animal communication, 2nd edn.Sinaeur, Sunderland

Brown J (1969) "Funky drummer." Funky drummer (Parts 1 and 2). King Records, Vinyl record

Brown J (1976) "Get Up offa that thing." Get up offa that thing. Universal/Polydor, Vinyl record

Brunswik E (1956) Perception and the representative design of psychological experiments. University of California Press, Berkeley

Cage J, Hiller L (1967–1969) HPSCHD. Champaign, Illinois

Clark H (1996) Using language. Cambridge University Press, Cambridge

Clutton-Brock T, Albon S (1979) The roaring of red deer and the evolution of honest advertising. Behavior 69:145–170

Cohen J (1962) Information theory and music. Behav Sci 7 (2):137–163

Collins N (2003) Generative music and laptop performance. Contemp Music Rev 22(4):67–79

Cone E (1974) The composer's voice. University of California Press Books, Berkeley

Cope D (1992) Computer modeling of musical intelligence in EMI. Comput Music J 16(2):69–83

Cope D (1996) Experiements in musical intelligence, vol. 12. A-R Editions, Madison

Cope D (1991) Recombinant music: using the computer to explore musical style. Computer 24(7):22–28

Cope D (2004) Virtual music. MIT Press, Cambridge

Cross I (2003) Music and evolution: consequences and causes. Contemp Music Rev 22(3):79–89

Cross I (2001) Music, mind and evolution. Psychol music 29(1):95–102

Crypton Future Media (31 August 2007) Hatsune Miku Product Page. <http://www.crypton.co.jp/mp/pages/prod/vocaloid/cv01.jsp>. Accessed 14 Dec 2012

Cumming N (2000) The sonic self: musical subjectivity and signification. Indiana University Press, Bloomington

Dibben N (2001) What do we hear when we hear music? Music perception and musical material. Music Sci 5(2):161–194

Dissanayake E (1995) Homo aestheticus: where art comes from and why. University of Washington Press, Seattle

Dissanayake E (2008) If music is the food of love, what about survival and reproductive success? Music Sci 12:169–195

Dissanayake E (1998) Sociobiology and the arts: problems and prospects. In: Baptist J, Cooke B (eds) Sociobiology and the arts. Bedaux, Amsterdam, 27–42

Dissanayake E (1988) What is art for? University of Washington Press, Seattle

Eisenberg E (2005) The recording angel: music, records, and culture from Aristotle to Zappa. Yale University Press, New Haven

Eno B (2005) Bottomliners: Another day on earth. Hannibal, Compact disc

Fastl H (2005) Psycho-acoustics and sound quality. In: Blauert J (ed) Communication acoustics. Springer, New York

Fitch W, Hauser M (2002) Unpacking 'Honesty': vertebrate vocal production and the evolution of acoustic signals. In: Simmons M, Fay R, Popper A (eds) Acoustic communication. Springer, New York, 65–137

Fitch WT (2006) The biology and evolution of music: a comparative perspective. Cognition 100(1):173–215

Fitch WT (2000) The evolution of speech: a comparative review. Trends Cogn Sci 4(7):258–267

Frenette B (2007) Daniel Lanois: in the studio. www.youtube.com. <http://www.youtube.com/wa tch?v=75dYhzQ7Is0&feature=related> Accessed 14 Oct 2007

Frith S (1998) Performing rites: on the value of popular music. Harvard University Press, Cambridge

Frith S (2004) Popular music: critical concepts in media and cultural studies. Routledge, London

Frith S (2012) The place of the producer in the discourse of rock. In: Frith S, Zagorski-Thomas S (eds). The art of record production. Ashgate Publishing Limited, Farnham, pp 207–221

Guggenheim D (2011) From the sky down. BBC Worldwide Digital video disc

Gabriel P (1993) XPLORA1: Peter Gabriel's secret world. XPLORA1: Peter Gabriel's Secret World. Real World Media

Grafen A (1990) Biological signals as handicaps. J Theor Biol 144:517–546

Hamlen W (1991) Superstardom in popular music: empirical evidence. Rev Econ Stat 73(4):729–733

Hammond K, Stewart T (2001) The essential Brunswik: beginnings, explications, applications. Oxford University Press, Oxford

Hansen B (2012) Song reader. McSweeney's, San Francisco

Hansen B (2013) songreader.net homepage. E&E, Eyes and Ears. <http://songreader.net/>. Accessed 10 June 2013

Hargreaves D, North A (1997) The social psychology of music. Oxford University Press, New York

Hargreaves D (2005) Musical communications. Oxford University Press, Oxford

Hedges S (1978) Dice music in the eighteenth century. Music Lett 59(2):180–187

Iyer V (2002) Embodied mind, situated cognition, and expressive microtiming in African-American music. Music Percept 3:387

Jackendoff R, Lerdahl F (2006) The capacity for music: what is it, and what's special about it? Cognition 100(1):33–72

Johnson-Laird PN (1991) Jazz improvisation: a theory at the computational level. In: Howell P, West R, Cross I (eds) Representing musical structure. Academic Press Inc, San Diego, pp 291–325

Juslin P, Laukka P (2003) Communication of emotions in vocal expression and music performance: different channels, same code? Psychol Bull 129(5):770–814

Juslin P, Scherer K (2008) Vocal expression of affect. In: Harrigan J, Rosenthal R, Scherer K (eds) The new handbook of methods in nonverbal behavior research. Oxford University Press, New York, pp 65–135

Juslin P, Sloboda J (2010) Handbook of music and emotion: theory, research, applications. Oxford University Press, Oxford

Justus T, Hutsler J (2005) Fundamental issures in the evoutionary psychology of music: assessing innateness and domain specificity. Music Percept 23(1):1–27

Katz M (2010) Capturing sound, how technology has changed music. University of California Press, Berkeley

Keil C (1987) Participatory discrepancies and the power of music. Cult Anthropol 2(3):275–283

Kendon A (1990) Conducting interaction: patterns of behavior in focused encounters. Cambridge University Press, Cambridge

Kendon A (2004) Gesture: visible action as utterance. Cambridge University Press, Cambridge

Kivy P (1989) Sound sentiment: an essay on the musical emotions, including the complete text of "The Corded Shell". Temple University Press, Philadelphia

Kramer L (1995) Classical music and postmodern knowledge. University of California Press, Berkeley

Kramer L (2001) Musical meaning: toward a critical history. University of California Press Books, Berkeley

Krebs J, Dawkins R (1984) Animal signals: mind-reading and manipulation. In: Krebs J, Davies N (eds) Behavioral ecology: an evolutionary approach, 2nd edn. Blackwell, Oxford, pp 380–402

Labov W (1973) Sociolinguistic patterns. University of Pennsylvania Press, Philadelphia

Laurel J (2006) Innateness, learning, and the difficulty of determining whether music is an evolutionary adaptation: a commentary on justus & hutsler (2005) and mcdermott & hauser (2005). Music Percept 24(1):105–109

LCD Sound System (2010) All i want: This is happening. EMI, Compact disc

Lefford MN (September 2004) The structure, perception and generation of musical patterns. Unpublished doctoral dissertation. Massachusetts Institute of Technology, Cambridge

Lerdahl F, Jackendoff RA (1996) Generative theory of tonal music. MIT Press, Cambridge

Letowski T (1989) Sound quality assessment: concepts and criteria. In: Audio Engineering Society 87th convention proceedings, 18–21 October 1989, p 2825

Lofting H (1920) The Story of Doctor Dolittle . New York, NY: Frederick A. Stokes

Longuet-Higgins H, Lee C (1984) The rhythmic interpretation of monophonic music. Music Percept 1(4):424–441

MacLarnon A, Hewitt G (2004) Increased breathing control: another factor in the evolution of human language. Evolut Anthropol 13:181–197

Marler P, Evans C, Hauser M (1992) Animal signals: motivational, referential, or both? In: Papoušek H, Jürgens U (eds) Nonverbal vocal communication: comparative and developmental approaches. Cambridge University Press, New York, pp 66–84

Marler P (1961) The logical analysis of animal communication. J Theor Biol 1:295–317

Mateas M (2005) Procedural literacy: educating the new media practitioner. On Horiz 13(2):101–111

Maynard-Smith J, Harper D (2003) Animal signals. Oxford University Press, Oxford

Mazzoni D, Dannenberg RB (2002) A fast data structure for disk-based audio editing. Comput Music J 26(2):62–76

McCartney J (2002) Rethinking the computer music language: SuperCollider. Comput Music J 26(4):61–68

McComb K, Taylor A, Wilson C, Charlton B (2009) The cry embedded within the purr. Curr Biol 19(13):R507–R508

McDermott J, Hauser MD (2005) The origins of music: innateness, uniqueness, and evolution. Music Percept 23(1):29–59

Meyer L (1957) Meaning in music and information theory. J Aesthet Art Crit 15(4):412–424

Meyer L. (1989) Style and music: theory, history, and ideology. University of Chicago Press, Chicago

Moorefield V (2005) The producer as composer. MIT Press, Cambridge

Narmour E (1990) The analysis and cognition of basic melodic structures: the implication-realization model. University of Chicago Press, Chicago

National Public Radio, Staff (2012a) Jack White on his own, tells other people's stories transcript. 1 June 2012. www.npr.org. Staff, Guy Raz National Public Radio. 5 October 2012 <http://www.npr.org/templates/transcript/transcript.php?storyId=154157412>.

National Public Radio, Staff (2012b) Jack White on his own, tells other people's stories. 1 June 2012. Guy, National Public Radio Raz. 5 October 2012 <http://www.npr.org/2012/06/03/154157412/jack-white-on-his-own-tells-other-peoples-stories>.

Orlarey Y, Graf A, Kersten S (2006) DSP Programming with Faust, Q and SuperCollider. 4th International Linux Audio Conference. Karlsruhe: ZKM | Zentrum fur Kunst und Medientechnologie 39–47

Osgood C, Walker E (1959) Motivation and language behavior: a content analysis of suicide notes. J Abnorm Soc Psychol 59(1):58–67

Patel AD (2006) Musical rhythm, linguistic rhythm, and human evolution. Music Percept 24(1):99–104

Paz EB, Ceccarelli M, Otero JE, Sanz JLM (2010) Medieval machines and mechanisms. In: Paz EB, Ceccarelli M, Otero JE, Sanz JM. A brief illustrated history of machines and mechanisms. Springer, Dordrecht, pp. 65–90

Puckette M (1991) Combining event and signal processing in the max graphical programming environment. Comput Music J 15(3):68–77

Puckette M (1996) Pure data. International Computer Music Conference. San Francisco: International Computer Music Association, pp. 224–227

Puckette M (1988) The patcher. International Computer Music Conference. San Francisco: International Computer Music Association, pp. 420–429

Ramachandran VS, Hirstein W (1999) The science of art: a neurological theory of aesthetic experience . J Conscious Stud 6(6–7):15–51

Reddy V (2008) How infants know minds. Harvard University Press, Cambridge

Repp B (1990) Patterns of expressive timing in performances of a Beethoven minuet by nineteen famous pianists. J Acoust Soc Am 88: 622–641

Robyn (2010) Hang with me: Body talk. Cherrytree Records/Cherrytree/Interscope/Interscope/Konichiwa, Compact disc

Scherer K (1995) Expression of emotion in voice and music. J Voice 9(3):235–248

Scherer K, Johnstone T, Klasmeyer G (2003) Vocal expression of emotion. In: Davidson E, Scherer K, Goldsmith H. (Eds) Handbook of affective sciences. Oxford University Press, New York, pp. 433–456

Schneck D, Berger D (2005) Music effect: music physiology and clinical applications. Jessica Kingsley Publishers, London

Schwanauer S, Levitt D (Eds) (1993) Machine modes of music. MIT Press, Cambridge

Seyfarth M, Cheney D (2003) Signalers and receivers in animal communication. Ann Rev Psychol 54:145–173

Shannon C, Weaver W (1949) The mathematical theory of communication, Vol 1. University of Illinois Press, Champaign

Smokey Robinson and the Miracles (1967) Tears of a Clown: Make it happen. Motown. Vinyl record.

Steedman M (1984) A generative grammar for jazz chord sequences. Music Percept 2(1):52–77

Temperley D (2006) Music and probability. MIT Press, Cambridge

Temple S (1936) Goodnight my love: Stowaway. 20th Century Fox, Motion picture.

ten Cate C et al (2009) Tinbergen revisited: a replication and extension of experiments on the beak colour preferences of herring gull chicks. Anim Behav 77(4):795–802

The Beach Boys (1966) Pet Sounds. Capitol Records, Vinyl record.

The Police (1983) Every breath you take: Synchronicity. A&M, Vinyl record.

Tinbergen N (1954) Curious naturalists. Basic Books, New York

Tinbergen N (1948) Dierkundeles in het meeuwenduin. De Levende Natuur 51:49–56

Toft R (2010) Hits and misses: crafting top 40 singles, 1963–1971. Continuum International Publishing, New York

Tomasello M, Call J (1997) Primate cognition. Oxford University Press, Oxford

Tomasello M (2008) Origins of human communication. MIT Press, Cambridge

Trevarthen C (2011) What is it like to be a person who knows nothing? Defining the active intersubjective mind of a newborn human being. Infant Child Dev 20:119–135

Tyrangiel J (2012) Auto-tune: why pop music sounds perfect. 5 February 2009. 23 November 2012 <http://www.time.com/time/magazine/article/0,9171,1877372-1,00.html>

U2 (1984) Pride in the Name of Love: The Unforgettable Fire. Island Records, Vinyl record

Unknown. Brian Eno on voices. (2006). www.youtube.com. <http://www.youtube.com/watch?v=HGB8M1Gh0Eo> Accessed 20 Jan 2013

Velleman S, Vihman M (2007) Phonology development in infancy and early childhood: implications for theories of language learning. Pennington M (ed) Phonology in context. Palgrave Macmillan, New York, pp. 25–50

Vercoe B, Ellis D (1990) Real-time CSound: software synthesis with sensing and control. International Computer Music Conference. Glasgow: International Computer Music Association. pp. 209–211

Wadhams W (2001) Inside the hits. Berklee Press, Boston

Wallin N, Merker B, Brown S (Eds) (2001) The origins of music. MIT Press, Cambridge

White J (2011) I guess I should go to sleep: Blunderbuss. Third Man/XL Recordings/Columbia, Compact disc

Wiley R (1983) The evolution of communication: information and manipulation. Halliday T, Slater P (eds) Communication. Blackwell, Oxford, pp. 157–189

Wittgenstein L (1953) Philosophical investigations. Blackwell, Oxford

Chapter 4
Mediated Interactions and Musical Expression—A Survey

Dennis Reidsma, Mustafa Radha and Anton Nijholt

4.1 Introduction

The dawn of the information and electronics age has had a significant impact on music. Digital music creation has become a popular alternative to playing classical instruments, and in its various forms has taken a place as full-fledged class of instrument in its own right. Research into technological or digital instruments for musical expression is a fascinating field which, among other things, tries to facilitate musicians and to improve the art of musical expression. Such instruments broaden the available forms of musical expression and provide new modes for expression, described by some as a reinvention of the musician's proprioceptive perception (Benthien 2002; Kerckhove 1993). They can make musical expression and musical collaboration more accessible to non-musicians and/or serve as educational tools. Technology can also eliminate the boundaries of space and time in musical collaboration or performance, or enhance it, by providing new channels of interaction between performers or between performer and audience. Furthermore, technology in itself can be a collaborating partner in the form of a creative agent, co-authoring, helping or teaching its user. Finally, Beilharz brings forward the human desire for post-humanism and cyborgism in musical expression as a goal in itself to explore mediating technologies (Beilharz 2011).

In this chapter we will survey music technology through various lenses, exploring the qualities of technological instruments as tools, media and agents and investigating the micro-coordination processes that occur in musical collaboration, with the long range goal of creating better technological artifacts for music expression.

D. Reidsma (✉) · M. Radha · A. Nijholt
Human Media Interaction, University of Twente, PO Box 217,
7500 AE, Enschede, The Netherlands
e-mail: d.reidsma@utwente.nl

M. Radha
e-mail: mustafa.radha@gmail.com

A. Nijholt
e-mail: a.nijholt@utwente.nl

N. Lee (ed.), *Digital Da Vinci,* DOI 10.1007/978-1-4939-0536-2_4,
© Springer Science+Business Media New York 2014

As a starting point, the next section discusses our theoretical starting point for looking at the relation between (technological) instruments and their users.

4.2 Users and Musical Instruments: Conceptual Framework

Musical instruments are designed with the purpose of engaging, intriguing and appealing to humans. There are various categories of user of musical instruments (most importantly, *musicians* and *non-musicians*) who differ in their needs and interactions with music instruments (Coletta et al. 2008; Akkersdijk 2012; Beaton 2010).

Understanding how humans deal with technology, including music technology, and how technological artifacts influence our world, are elements of an ever-growing body of research. Verplank (2011) suggests to use a *tool-media-agent* model to understand the impact of (music) technology. Verbeek, building upon Ihde's post-phenomenological work (Ihde 1986), proposes three manifestations of technology: as a tool for a human; as a medium between two humans; or as an agent in itself, part of the social network (2005). An important aspect in each of these manifestations of human music interaction is the underlying *collaboration process*. Tatar (Lee et al. 2012; Tatar 2012) urges exploration of micro-level, situated actions and their broader outcomes, a process she coins as *micro-coordination*. Studying micro-coordination in interaction settings gives the designer a better understanding of the collaborative phenomenon and, as a consequence, better tools to design technology for the domain. In the music making domain, insight into this topic will enable us to invent better (mediating) instruments by accounting for the necessary micro-co-ordination, and to implement realistic music-creating agents to work with humans that are able to understand—and participate in—the necessary processes of coordination. We apply these models to musical instruments as follows.

Tools—Music instruments as tools merely serve as a means for their user to more easily achieve a goal. In Sect. 3, we will investigate the design of these tools in the light of different modes of interaction and how they satisfy various user needs. How can technology be used to enhance and create new interfaces for musical expression? Which technological modes of interaction can be exploited for musical expression? How can these modes of interactions serve the needs of different types of users?

Media—Instruments as media provide a channel of communication with other humans. Instruments of this type are mediators in social (human to human) interaction. In Sect. 4, we will look at how communication related to the context of collaborative musical expression can be realized. How can instruments capture the micro-coordination involved in collaborative musical expression? What micro-coordinating communication occurs in collaborative musical expression? How can technology be a mediating channel for this micro-coordination? How can technology augment the taxonomy of micro-coordination, enabling new ways of coordination?

Agents—An agent is an artifact possessing agency: it has beliefs, desires and intentions (Georgeff 1998), is part of the social structure as an entity in itself, and thus introduces the field of human-agent or human-robot interaction. We will study the design prerequisites for co-creative musical agents in Sect. 5.

4.3 Interaction Modes for Musical Expression: The Instrument as Tool

In this section, the technological instrument for musical expression will be discussed as a tool, investigating how new interaction modes enable new forms of musical expression. We will discuss different novel technological interaction modalities and how these complement the musician's needs, and look at how technology can support *non-musicians* in achieving music making experiences.

4.3.1 Technology for Multimodal Music Interaction

Technological innovations allow for interaction modalities that go far beyond the simple keys of a digital piano. This section discusses three of the most important new modalities: tactile interfaces, gestural control, and brain-computer interaction for music making.

Tactile Interfaces

Many technological music instruments rely on touch, tangibility and haptics as control interfaces. The success of the tactile interface can be contributed to several factors.

The first reason is the ease of use or intuitiveness. Tactile interfaces exploit the user's natural affinities with vision and touch. We are used to explore the world through our eyes and manipulate it with our hands. Making this very real form of interaction with the world into an interface through augmented reality gives us many possibilities as is shown in many diverse projects (Fikkert et al. 2009; Jorda et al. 2007; Levin 2006; Patten 2002; Poupyrev et al. 2000; Raffa 2011).

Haptic interfaces may also invite the user into new forms of expression. Bill Verplank (Verplank et al. 2002) has done research into the expressive effects of haptic force feedback as a means of informing the user of their action. The power of haptics is illustrated in his *Plank* device, a slider with natural force feedback that resembles normality (the further you push the slider, the more resistance it produces). Such sophisticated interaction methods can be a powerful enhancement to the currently popular and successful touch-based and tangible instruments.

Another benefit is the extendable size of such interfaces. Many inventions like the FeelSound (Fikkert et al. 2009) and reacTable (Jorda et al. 2007) are table-sized

Fig. 4.1 The reacTable interface (Jorda et al. 2007)

tactile interfaces designed to enable multiple musicians to work together on the same piece.

Not only the usability, but also the utility of the instruments can vary widely. While most tactile music interfaces look the same, their purposes are often different because of the adaptable nature of tactile interfaces. In illustration, the idea of a table-sized touch screen with tangible objects on them has been reused for composition (Fikkert et al. 2009), spectrograph manipulation (Levin 2006), rhythm creation (Raffa 2011), mixing (Poupyrev et al. 2000; Jorda et al. 2007) and harmony creation (Jorda and Alonso 2006). This promises the possibility of an all-round tactile table-top installation for musical expression in many forms.

The last aspect which especially sets apart tactile interfaces from the old-fashioned MIDI knob/button controller for the laptop, is the appealing interface for the audience. Paine (2009) emphasizes the need for interfaces that give more insight into the creative effort of the musician. He states that the laptop musician creates a distance barrier between him and the audience since his actions are not apparent for the audience. Projects like the reacTable (Jorda et al. 2007) (Fig. 4.1) serve to make the performance of the electronic musician more appealing.

Gesture Interfaces

As computer vision and the detection of movement advance, gesture interfaces are more and more explored as a new mode of musical interaction. Gestures can be *quantified* to control parameters of the music, but can also be *sonified* directly into sound.

These two extremes can be generalized into a certain dimension of control mechanism design, *directness of control*. With indirect control, the output is only modified by the input. With direct control, there is a direct mapping of input to output.

An example of indirect control is the Mappe per Affetti Erranti (Coletta et al. 2008). The system monitors the movements of dancers on a stage and quantifies them into parameters that are applied to a preprogrammed music piece. The goal of the project is to let dancers conduct the music instead of the other way around. Virtual orchestra conducting interfaces (Borchers et al. 2004) and virtual conductors for real orchestras (Reidsma et al. 2008) share a similar mechanism. A non-gesture example of indirect control is the Wayfaring Swarms installation (Choi and Bargar 2011), a tabletop installation with music-generating swarms with which the user can interact to some extent.

Direct control gesture interfaces, in which the gestures are directly sonified, are scarce. A well-known realization of gesture sonification is the Theremin. According to Ward et al. (Ward and O'Modhrain 2008), the mastery of such instrument lies in the execution of managed movements. Goina and Polotti (2008) investigate how gestures can be sonified in general by identifying the elementary components of gestures, working towards the idea of melting the virtuosities of dance and music into a single art, which confirms the importance of managed movement.

Brain-Computer Interfaces

One of the latest additions to the repertoire of multimodal interaction technologies for HCI is that of Brain-Computer Interfacing: brain signals from the user are registered using, e.g., an EEG device, and signal processing techniques are used to obtain relevant information about the user's mental state (Nijholt et al. 2008). In its most primitive form, such an interface simply serves as a trigger which is activated through a specific kind of brain activity rather than through pushing a button. This allows musical expression to become available for people with physical disabilities that prevent them from playing regular instruments (Miranda 2006; Chew and Caspary 2011; Le Groux et al. 2010).

BCI interfaces have also been used for *direct control* of musical expression. Examples of this are the early work "Music for Solo Performer" by Lucier (Teitelbaum 1976) and "Staalhemel" (De Boeck 2011)—an installation in which a person walks under steel plates hung from the ceiling, and their brain signals are translated to small hammers that tap the steel plates according to the amplitudes of their brain signals. BCI based musical expression is also possible in multi-user settings. Sänger et al. (2012) looked at how collaboratively making music causes the brains of the musicians to synchronize. MoodMixer (Leslie and Mullen 2011) offers an EEG-based collaborative sonification of brain signals. Interestingly, these and similar project blur the distinction between direct control and indirect control. Consciously manipulating your own brain activity is possible, one can see what the effects are, relax, concentrate, think of something sad, think of something happy, etcetera. At the same time, brain activity is always there, it cannot be suppressed, and it can be manipulated in a feedback loop from the outside by exposing the user of the instrument to changes in the environment. Then, sonification is 'controlled' by what the user experiences, but the user experience is simultaneously strongly controlled by the output of the sonification.

4.3.2   Music Interaction for Non-Musicians

These new interface technologies not only serve musicians, but are also usable for non-musicians and people who are still learning to play music. Two keys to engaging non-musicians are (1) making it easy and intuitive to make music and (2) giving the non-musician a taste of musical achievement.

Table-top installations have already been discussed as intuitive and natural interaction methods for musical expression. They therefore can appeal more to non-musicians than conventional instruments. Efforts in tangible interfaces focusing on teaching music are the MIROR project (MIROR 2012) and Melody-Morph (Rosenbaum 2008). Both projects exploit the loose building blocks of the tangible interface to make a *configurable interface*. This eliminates the learning curve associated with figuring out how the interface responds to the user's actions, since the user can himself design the triggers and responses.

When technological instruments are used to offer the experience of a high level of musical performance to non-musicians without them (yet) having the real skills for such performance, it is called *simulated musical achievement*. An example of simulation of musical achievement are the Guitar Hero games (Harmonix Music Systems 2005). Miller (2009) explains the success of these games in the fact that they bear in them a form of *schizophrenic performance*, where the player, while not playing the instrument, does have the feeling of actually making the produced music. This separation between input and output (thus: schizophrenia) can essentially be generalized to the indirect control mechanisms discussed earlier. The levels of difficulty in the game represent the spectrum of directness. This principle can be applied to other interfaces such as gesture interfaces, where the interface can gradually transform from gesture parametrization into real sonification. Simulated musical achievement may in this way be used to motivate people to learn the actual skills for music making.

4.4   Mediated Communication in Musical Expression: The Instrument as Medium

A significant part of the research area concerned with musical expression focuses on the *collaborative* aspects of music-making. Many studies on this topic aim towards instruments that enable interesting forms of collaboration and interaction with other humans. In this section, we will look at research done to enable mediated human-human interaction in musical collaboration.

Communication in musical expression			
Micro-coordination in collaboration		**Performer-audience communication**	
Synchronization	*Differentiation*	*Performance facilitation*	*Interaction*
playing style	leader/follower role establishment	virtuosity	entrainment
-loudness	leading	affective expression	
-pitch	-pitch change	dance	
-rhythm	-loudness change	speech	
meta-information (spatial dislocation)	-rhythm change		
-linking sounds to their creators	following		
meta-information (temporal dislocation)	-following up on pitch	**Used channels**	
-awareness of action	-following up on rhythm		
-annotation	-following up on loudness	**visual channel**- non-verbal behaviour	
-shared representation		**verbal channel**- verbal behaviour	
-mutual modifiability		**auditory channel**- musical behaviour	

Fig. 4.2 Taxonomy of interactions in musical expression as found in literature survey

During musical collaboration, traditionally there are three channels present, namely the visual, the auditory and the verbal channels (Beaton et al. 2010). The visual channel deals with body movements, the auditory are actions performed with the instrument (e.g. a drumming break) and the verbal channel contains only utterances and words.

The actions performed over these channels can also fall into different categories based on their intentional goals, forming a taxonomy of actions over two dimensions: channels and goals. As an analogue, Bianchi-Berthouze (2013) has developed a classification of the movements performed in gaming, which consist of task-controlling, task-facilitating, affective, role-facilitating and social movements. We will establish such a taxonomy in this section. The complete taxonomy as explained in this section can be seen in Fig. 4.2.

We first divide actions into two contexts: collaboration between musicians (micro-coordination) and performance towards the audience. Micro-coordination between musicians can happen for three different collaborative goals: the first is to achieve mutual synchronization between the musicians (e.g. entrainment, mutual awareness of meta-information, countdown) and the second is to dynamically change the music in a coordinated manner by establishing roles (leading, following). Between the musician and the audience, there are "monologue" communications by the musician, part of the performance, and interactive communications where the musician and audience work together to make the event.

4.4.1 Collaborative Micro-Coordination

This section concerns the micro-coordination that happens in cooperative music-making. We discriminate between two types of micro-coordination in musical expression: the first has a goal of synchronizing the collaborators with each other, while the second type has an opposite goal: the goal of differentiating the music to create something new and to ensure a dynamic flow.

Synchronization

Synchronization is done throughout the whole collaboration: it is used to initiate the music, to keep the different musicians in harmony and rhythm throughout the collaboration and finally to finish the created musical piece in a synchronized manner. Synchronization can occur over all 3 channels of communication.

Before starting musical expression, musicians have to communicate which piece of music to play (usually verbal) and perform a countdown of some sorts before commencing. This countdown can happen either verbally by counting, auditory by, for example, 4 hits on a percussion instrument or visually, for example by the leader through an exaggerated body movement before starting.

During musical expression, synchronization is needed constantly to work together. It appears that not only hearing each other playing music, but also being able to see each other while doing so, are important for synchronization. For example, Beaton (Beaton et al. 2010) found out that musicians could collaborate better with humans than with robotic performers because robotic performers lack visual feedback. Akkersdijk (2012) showed that the placement of a screen between musicians prevented them from synchronizing as well as they would when they can see each other.

The robotic improvisational Marimba-playing agent Shimon (Hoffman and Weinberg 2010) was designed based on research into meaningful movements for robots (Hoffman and Breazeal 2006, 2008; Hoffman et al. 2008) in order to emulate a human musician. Unique to this project is the effort put in the design of movements. The robotic arms that play the Marimba make elaborate movements to express several properties of the music: the motion size naturally denotes the loudness of a keystroke; the location of the gesture naturally corresponds to the pitch of the keystroke; the movements are exaggerated for further visibility for the other musicians. In another project, Varni et al. (2008) showed that humans collaborating with each other try to entrain their body movements to the external jointly established rhythm, and that this becomes more difficult when the humans are not able to see each other.

Synchronization During Spatial Dislocation

The ten-hand piano (Barbosa 2008), part of the Public Sound Objects (PSO) project (Barbosa and Kaltenbrunner 2002), is envisioned as a piano playable by multiple people who do not necessarily see each other, in public spaces. The user interface represents different users as circles on the screen. The color of these circles denotes the pitch and the size denotes the loudness of the music produced by that person. This way, users who cannot see each other still can differentiate the mix of produced sounds in terms of their origin. In other words: this provides necessary meta-information about the joint musical product, which is necessary for mutual collaboration. This helps synchronization as users are able to relate sounds to the persons producing them.

Synchronization During Temporal Dislocation

Bryan-Kinns (Bryan-Kinns and Hamilton 2009; Bryan-Kinns 2004) has identified more meta-information in his studies on mutually engaging music interaction in dislocated setting. He highlights the following properties that are needed to communicate the necessary meta-information for interesting collaboration when both spatial and temporal co-location are not possible.

- **The mutual awareness of action**—highlighting new contributions to the joint product and indicating authorship of components
- **Annotation**—being able to communicate in and around a shared product, and being able to refer to parts of the product helps participants engage with each other
- **Shared and consistent representation**—the participants find it easier to understand the state of the joint product and the effects of their contributions if the representations are shared and consistent
- **Mutual modifiability**—editing each other's contributions increases mutual engagement

In the Daisyphone project (Bryan-Kinns 2012) Bryan-Kinns has implemented these properties into the interface, which can be seen in Fig. 4.3. Since the interface is designed with these principles in mind, it allows both temporal and spatial dislocation during a collaboration.

Coordinating Differentiation

While synchronization is an essential part of joint musical expression, it solely serves the purpose of aligning the expressions of the musicians to each other. In order to make the collaboration dynamic and create new expressions, other tools are used. An established interaction to facilitate this differentiation is the leader/follower dynamic. During the collaboration, a leader will be communicated who can improvise something new into the collaboration. The follower(s) must be able to adapt to the leader's new piece. The leadership role can also be reassigned during the collaboration in order to give multiple collaborators the chance to improvise.

Beaton (Beaton et al. 2010) has established that leadership is communicated over different channels, being the visual, verbal and auditory ones. Reidsma et al. (2008) show that for a virtual music conductor (see Fig. 4.4) to work, it has to take into account what the orchestra is doing (ie. it is a two-way process between leader and follower). Borchers et al. (2004) have shown the same for the inverse case: a virtual

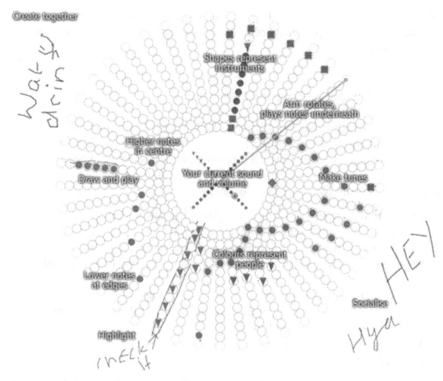

Fig. 4.3 User-interface of the Daisyphone (Bryan-Kinns 2012) with annotated representation and hand-written textual communication

Fig. 4.4 Bidirectional interaction with a virtual conductor (Reidsma et al. 2008)

orchestra conducted by a real conductor. This idea is also confirmed by Akkersdijk (2012): the leaders in her experiment used eye contact to monitor whether the follower was actually still following. Akkersdijk also identified several actions that leaders employ to communicate their intentions. Leaders entrain their breathing, head nods, body movements and arm movements with the rhythm while followers employ mimicry (most in arms, least in head) to show that they are picking up on the proposed rhythm. This is in parallel to the work by Varni et al. (2008) about emotional entrainment to an external rhythm. It could be that while entrainment is used to maintain a rhythm it is also used to change rhythm by the leader. The follower's feedback in the form of mimicry is used by the leader as a measure of how well his improvisation is catching on to his/her partners.

4.4.2   Performer-Audience Relation

A different relation is the one between the performer and the audience. While the communication from the musician to the audience is straightforward, being the performance as a whole, the feedback from the audience is a little complicated yet important for the musician to be aware of his/her audience as well. We look at the monologue communication from musician to audience, which we shall call the *Performance* and the dialogue between musician and audience which completes the event, which we will term the *Performance-related interaction*.

Performance

Besides the created music, several other channels are often employed by the musician in a performance. Munoz (2007) emphasizes the importance of these channels, coining movement as the motor of sound and intention as the impulse of gesture, which leads to the inevitable relation between intentional body movements and musical expression. We found 4 different categories of performance: virtuosity, affective expressions, dance and speech. The Marimba robot (Hoffman and Weinberg 2010) was deliberately designed with only 4 arms to operate a complete Marimba, in contrast with many robotic performance projects that assign an arm to each key on an instrument like the MahaDeviBot (Eigenfeldt and Kapur 2008). This was done to express virtuosity, since the fast movements of a limited number of operating arms to generate the music is perceived as a form of mastery over the instrument. This is further emphasized by Paine (2009) who notes the fast bodily movements as an important factor in the appreciation of a performance by the audience. Musicians employ affective expressions to tell something about and amplify the power of the conveyed emotions in the music. Thompson et al. (2008) show that body movements and facial expressions are often used for this effect. Mobile instruments allow the musician to perform limited dance while playing the instrument. Instruments such as the Mappe per Affetti Erranti (Coletta et al. 2008) envision to exploit new interfaces to enable a complete blend of dance and music virtuosity. Verbal expressions are also used to introduce a piece of music or for affective amplification of the music.

Performance-Related Interaction with the Audience

Tarumi et al. (2012) have informally explored the communication between performer and audience and have found several gestures employed by the audience:
hand waving, hand clapping, hand joggling, pushing hands in the air and towel
swinging. In addition, the audience also uses words, either short words like "yeah!"
and "oh!" or phrases of a song altogether. All these action share that they are a form
of entrainment to the music of the musician, expressing compulsion and engagement. Takahashi et al. (2011) have built hand-clapping robots, placed in front of
a streaming performer (performing over an on-line stream) and operated by his
remote audience. Visitors of the performance on-line can use buttons on the website
to make the robots clap to the musician, enabling hand clapping while being spatially dislocated.

4.5   Co-Creative Agents: The Instrument as Agent

Robotic music-making is an emerging field within music technology. We want to
look at the collaborative aspects and the interaction with real musicians in the co-
creative agent, being an agent with the ability to collaborate with other artificial or
human musicians. The preconditions for such an agent are its abilities to perceive
and create music and to understand and act in social contexts. We will shortly look
at music perception and music creation and agents in a social context (ie. a co-
creative agent).

4.5.1   Music Perception

Before an agent can be part of a musical ensemble, it must, trivially, be able to make
music and listen to the music of its partners. Music perception is the topic of perceiving music and has received a significant amount of attention in research, since
its applications vary across a wide array of music-related settings. McDermott and
Oxenham (2008) have reviewed recent developments in the field. They characterize
the main features of music (e.g. pitch, timbre and rhythm) and point out the cultural
differences in the way music is perceived. In their study, they elect cognitive and
neurosciences as fields where music perception can benefit from. Modelling music
perception in a cognitive manner is effective in extracting affective features from
music. State-of-the-art examples of systems for the perception of music in a musical agent are Varni's system for the analysis of interaction in music (Varni et al.
2010), discussed more thoroughly in the next part about social awareness and robotic drummers that are able to collaborate on rhythmic dimensions like the robots
built by Weinberg and Driscoll (2006) and Crick et al. (2006).

Hoffmann and Weinberg's *Shimon* (2010) reduces the problem of harmony perception to simple pitch detection. Instead of working with a note representation of what it hears, it just listens to the pitch of the music and moves its robotic arms to the approximate location of corresponding pitch locations on the Marimba. The anticipatory system in the Marimba furthermore builds patterns of how its partner progresses through harmony to anticipate where the arms should be, making it possible to play along without delay.

4.5.2 Music Creation

Meta-creation involves using tools and techniques from artificial intelligence, artificial life, and machine learning, themselves often inspired by cognitive and life sciences, to create music. Musical meta-creation suggests exciting new opportunities to enter creative music making: discovery and exploration of novel musical styles and content, collaboration between human performers and creative software "partners", and design of systems in gaming and entertainment that dynamically generate or modify music. We have identified three general approaches in musical agents for meta-creation, being the model-driven, the data-driven and the cognitive approaches. We also look at a specific application of meta-creation for improvisation: continuation.

The **model-driven** approach models music in terms of aesthetic and organized rhythm and harmony. This model of music is then used by the agent to produce music that fit the model. Eigenfeldt et al. have developed Harmonic Progression (Eigenfeldt 2009). Another example is the Kinetic engine, a real-time generative system (Eigenfeldt 2010) which has been used in different contexts, amongst which improvising ensembles (e.g. MahaDeviBot (Eigenfeldt and Kapur 2008)).

Another approach to music creation is the **data-driven** approach. This strategy generates music from a pool of existing compositions. The idea is implemented in MusicDb (Maxwell and Eigenfeldt 2008), which consists of a music information database and an accompanying query system, which both form a streamlined algorithmic composition engine.

The **cognitive learning** method models a music creating agent to imitate human-like musical creativity. It is concerned with the learning process: what features of music do we store and how do we store them in our minds? The hierarchical memory theory (Lerdahl and Jackendoff 1996) defines a hierarchical storage, while the Long and Short Term Memory approach (Eck and Schmidhuber 2002) emphasizes the importance of temporal features in this hierarchy. Both theories are applied in the Hierarchial Sequential Memory for Music (Maxwell et al. 2009).

The Sony Continuator (Pachet 2002) is an example of **continuation**, a special form of meta-creation in which a machine takes up the music after the human player stops, while maintaining the style initiated by the human performer. By employing Monte Carlo Markov chains (Gamerman and Lopes 2006), the Continuator learns the music style of a performer and then continues in the same style, providing a partner during improvisation.

4.5.3   Social Awareness and Behavior

A part of co-creation that is especially interesting in the light of mediated interaction is social behavior and awareness in co-creative agents. A co-creative agent must be able to understand and act out the multi-modal communication described in Sect. 4. We discuss a few systems on social awareness and then proceed to highlight a few projects on social behavior in collaborative settings. We will also shortly visit choreography as a special category of social behavior.

Social Awareness

For an autonomous musical agent to function in a collaborative setting, it needs to be aware of that setting. A system for the analysis of social interaction in music has been developed by Varni et al. (2010). The key factors of their system are driving interpersonal synchronization of participants in a music ensemble, the identification of roles (e.g., leadership, hierarchy), the general principles of how individuals influence each other and the factors that drive group cohesion and the sense of shared meaning. This system shows promises for the development of social awareness in collaborative agents for musical expression.

Social Behavior

Social behavior is the acting out of the social identity of the agent in a collaboration. The agent must be able to act accordingly to social rules. Communication between musicians was studied in Sect. 4. This information is useful to implement social agents.

Shimon (Hoffman and Weinberg 2010) was already provided as an example of a social musical collaborator. It contains several interaction modules to make music in a social way. Its *call-and-response* system responds to a musical phrase with a chord sequence. Its *opportunistic overlay improvisation* tries to anticipate upcoming chords by its partners and play notes within that chord. The *rhythmic phrase-matching improvisation* is a decaying-history probability distribution for rhythm, classifying which rhythm to use to stay synchronized with its partner.

If we split up roles in *following* and *leading*, knowledge from the related field of conductor simulation can be used as a starting point. Borchers et al. (2004) have developed the Personal Orchestra, a conducting environment in which the user can conduct a virtual orchestra. The conducting behavior of the user is translated into musical parameters (e.g. tempo) and the virtual orchestra adjusts how they play a (predefined) music piece according to those parameters. Combining the extraction of parameters with Eigenfeldt & Pasquier's Harmonic Progression (discussed above), we can let a model-driven creative agent be influenced by the user directly. The MIROR project (MIROR 2012) with its focus on reflexive music generation can also serve well in such a situation. Their installation is able to react upon the user's input without interrupting the music.

The counterpart of a following role is the leading role. Beaton et al., in their study that compares experienced musicians with inexperienced people (Beaton et al. 2010) in the context of human-computer musical collaboration, report that inexperienced musicians perceived a leading agent as helpful to guide them through the process of making music. This contrasts with their findings on experienced musicians in the same setting: they tend to prefer a following musician. This means that when the intended user category of an autonomous music creating agent is the non-musician, a leading role is favorable. To implement leading properties, again we can look at research on conducting systems. In this case, we look at a virtual conductor that leads a human orchestra made by Reidsma et al. (Raffa 2011). It is pointed out that to successfully lead in a musical setting, one needs to communicate intentions and work well together with the human musicians. This is why special attention is paid to the subtle dynamic of leading and following when fulfilling the role of a leader.

In Shimon, an artificial head is used to communicate leadership: when the robot looks at the musician, it wants him to improvise (and will follow the musician). If it looks at its Marimba, it will improvise itself (lead).

Choreography

Choreography is especially interesting when the robotic musician is performing for an audience. As we have shown in Sect. 4, the expression of virtuosity, affect and even dance through the non-verbal channel are important for an interesting performance. It can also serve as a way of communication, using realistic movements when playing an instrument to give other musicians a way to anticipate the robot's musical phrases. Liveness has also been noted in several studies as an important factor for a performing robot, both for an audience and for its partners (Hoffman and Breazeal 2008). A choreographic system is needed for the robot to be interesting in a performance and pleasurable as a music partner.

4.6   Discussion

In this survey, we have established a broad view of the state-of-the-art in instruments for music expression. Instrument design principles have been categorized in the tool, media and agents paradigms as well as into different user categories, of which the main distinction is between musicians and non-musicians.

When viewing instruments as tool, the instrument is said to be a thing that the human user can use to interact with the physical world. It is important to consider the user groups that the instrument is being designed for. The interface can be tangible, which is a preferred interface for musicians as tangibility allows natural feedback to occur in the form of haptics. Tangibility is preceded by the touch-interface and followed up by haptics in terms of interaction sophistication. Other interfaces can be gesture-based, imagery-based or iconic and touch-oriented. The type of interface can have a strong influence on the control mechanism. Whereas tangible interfaces provide a robust control mechanism that enables direct control, it is very hard to have direct control with gesture-based controls, although there have been successful attempts and studies are ongoing on the topic of gesture sonification. Not only the interface but the exterior design is also an important determinant for the reception of an instrument. The use of computer interfaces can suffer the lack of allure that is present in performances on traditional instruments. When designing for a specific type of user, the exterior design should also be adapted towards the expectation of that user group.

Instruments can be designed as media, which means that they transcend the concept of a tool. Media enable the channeling of communication between different humans as part of their design and thus enable interesting new forms of musical collaboration. We distinguished between spatial and temporal co-location in a collaboration. Usually, musicians are both spatially and temporally co-located when performing music with tools. With media, we are able to remove one or both forms of co-location to conceive new forms of musical collaboration. When designing a spatially de-located instrument, it is important to make sure that the communication that is present otherwise is substituted in some manner. A definer of success is to provide a means for the user to distinguish between the tunes made by the different collaboration partners. Definers of success in temporally de-located collaborations are the mutual awareness of action, annotation capabilities, shared and consistent representations and mutual modifiability. The possibilities of musical instruments as media are huge, promising flexibility, new forms of social interaction and user-generated artistic content in public areas.

When designing agents for musical expression, a main issue is the design of computational creativity. There are three main categories of computational musical creativity. The first is the model-driven approach, in which the designer can model the generation of aesthetically pleasing music and feed that model to the agent. Main studies have been concerned with the modelling of rhythm and harmonics. Secondly, we have the data-driven approach, which is the art of building an agent that is able to combine different existing musical pieces in seamless transitions in an interesting way. The last approach is cognitive learning, in which we try to model an agent's musical creativity learning process as humans do so. Key concepts are long and short term memory, hierarchical memory and neural network methods.

We are not only interested in robotic music authorship, which is the area of computational creativity, but also in the possibilities of a virtual musician augmenting a human in his musical expressions. This is when an agent can serve as a smart instrument. Next to computational creativity, one of the things we also have to account for are social awareness, which primarily consists of role recognition and role execution within a musical ensemble. There are promising systems for the analysis of social context within musical ensembles. Theory on the execution of musical roles can be lent from the domain of musical conductor research. Studies have shown that non-experienced users favor different social behavior from the agent than experienced musicians. The last important piece to enable autonomous agents as part of an instrument is music perception. The agent must have the ability to experience the music of the human musician and for this, we need to implement methods of music perception. It has been suggested that the fields of cognitive and neurosciences can teach us about the cognitive methods of music perception.

In this chapter, we discussed various aspects of coordination in technology enhanced musical expression. The next step would be to better understand the interactive element of micro-coordination in collaborative musical expression, to ultimately build prototypes that can either substitute traditional communication channels or augment the interactive landscape in micro-coordination. Such research would target questions such as: What actions exist in the interactive landscape in collaborative musical expression (continuing on findings presented in this survey)? How can technology be used without disrupting this landscape? How can we use technology to enrich these interactions? We are looking at a field that, although already well-explored by many researchers, is still full of exciting possibilities for radically new interactive experiences.

References

Akkersdijk S (2012) Synchronized clapping: a two-way synchronized process. Capita Selecta paper

Barbosa Á (2008) Ten-hand piano: a networked music installation. In: Proceedings of the 2008 Conference on New Interfaces for Musical Expression, NIME '08, Genova, Italy, pp 9–12

Barbosa Á, Kaltenbrunner M (2002) Public sound objects: a shared musical space on the web. In: Proceedings of the First International Symposium on Cyber Worlds, CW '02, Washington, DC, USA, pp 9–11

Beaton B, Harrison S, Tatar D (2010) Digital drumming: a study of co-located, highly coordinated, dyadic collaboration. In: Proceedings of the 28th international conference on Human factors in computing systems, CHI '10, New York, USA, pp 1417–1426

Beilharz K (2011) Tele-touch embodied controllers: posthuman gestural interaction in music performance. Social Semiotics, Vol 21, issue 4. (Published Online), pp 547–568

Benthien C (2002) Skin: on the cultural border between self and the world. Columbia University Press. New York, USA

Bianchi-Berthouze N (2013) Understanding the role of body movement in player engagement. Hum Comput Int 28(1):40–75. (Published Online)

Borchers J, Lee E, Samminger W, Muhlhauser M (2004) Personal orchestra: a real-time audio/video system for interactive conducting. ACM Multimedia Systems Journal Special Issue on Multimedia Software Engineering, vol 9, issue 5, San Jose, USA, pp 458–465

Bryan-Kinns N (2004) Daisyphone: The design and impact of a novel environment for remote group music improvisation. In Proceedings of the 5th Conference on Designing Interactive Systems: processes, practices, methods, and techniques, DIS 2004, pp 135–144

Bryan-Kinns N (2012) Mutual engagement in social music making. In: Lecture Notes of the Institute for Computer Sciences, Social Informatics and Telecommunications Engineering, LNICST 78, pp 260–266. (Published Online)

Bryan-Kinns N, Hamilton F (2009) Identifying mutual engagement. Behav Inform Tech 31(2):101–125

Chew YCD, Caspary E (2011) MusEEGk: a brain computer musical interface. In: CHI '11 extended abstracts on human factors in computing systems. New York, NY, USA: ACM, pp 1417–1422. (A BCI example is MusEEGk)

Choi I, Bargar R (2011) A playable evolutionary interface for performance and social engagement. INTETAIN, vol 78. Genova, Italy, pp 170–182

Coletta P, Mazzarino B, Camurri A, Canepa C, Volpe G (2008) Mappe per Affetti Erranti: a multimodal system for social active listening and expressive performance. In: Proceedings of the 2008 Conference on New Interfaces for Musical Expression, NIME '08, Genova, Italy, pp 134–139

Crick C, Munz M, Scassellati B (2006) Synchronization in social tasks: Robotic drumming. In: Proceedings of the 15th IEEE International Symposium on Robot and Human Interactive Communication. ROMAN, Vol 15, Hatfield, UK, pp 97–102

De Boeck C (2011) Staalhemel: responsive environment for brainwaves. Available: http://www.staalhemel.com/

Eck D, Schmidhuber J (2002) Finding Temporal Structure in Music: Blues Improvisation with LSTM Recurrent Networks. In: Neural Networks for Signal Processing XII, vol 12. Martigny, Valais, Switzerland, pp 747–756

Eigenfeldt A (2009) A realtime generative music system using autonomous melody, harmony, and rhythm agents. In: XIII Internationale Conference on Generative Arts, Milan, Italy

Eigenfeldt A (2010) Realtime generation of harmonic progressions using controlled markov selection. In: Proceedings of the First International Conference on Computational Creativity, ICCC 2010. Lisbon, Portugal

Eigenfeldt A, Kapur A (2008) An agent-based system for robotic musical performance. In: Proceedings of the 2008 Conference on New Interfaces for Musical Expression, NIME '08, Genova, Italy, pp 144–149

Fikkert FW, Hakvoort MC, van der Vet PE, Nijholt A (2009) Feelsound: interactive acoustic music making. In: Proceedings of the International Conference on Advances in Computer Entertainment Technology, ACE2009, New York, pp 449–449

Gamerman D, Lopes HF (2006) Markov chain Monte Carlo: stochastic simulation for Bayesian inference, vol 68. Chapman & Hall/CRC

Georgeff M, Pell B, Pollack M, Tambe M, Wooldridge M (1998) The belief-desire-intention model of agency. Intelligent Agents V: Agents Theories, Architectures, and Languages. Paris, pp 1–10

Goina M, Polotti P (2008) Elementary gestalts for gesture sonification. In: Proceedings of the 2008 Conference on New Interfaces for Musical Expression, NIME '08, Genova, Italy, pp 150–153

Harmonix Music Systems (2005) Guitar hero video game series. DVD by RedOctane

Hoffman G, Breazeal C (2006) Robotic partners' bodies and minds: an embodied approach to fluid human-robot collaboration. Cognitive Robotics

Hoffman G, Breazeal C (2008) Anticipatory perceptual simulation for human-robot joint practice: theory and application study. In Proceedings of the 23rd national conference on Artificial intelligence, Chicago, IL, USA, pp 1357–1362

Hoffman G, Weinberg G (2010) Shimon: an interactive improvisational robotic marimba player. In: Proceedings of the 28th of the international conference extended abstracts on Human factors in computing systems. ACM, pp 3097–3102

Hoffman G, Kubat R, Breazeal C (2008) A hybrid control system for puppeteering a live robotic stage actor. In: Robot and Human Interactive Communication, 2008. RO-MAN 2008. The 17th IEEE International Symposium on. IEEE, pp 354–359

Ihde D (1986) Experimental phenomenology: an introduction. SUNY Press

Jorda S, Alonso M (2006) Mary had a little scoreTable* or the reacTable* goes melodic. In: Proceedings of the 2006 conference on New interfaces for musical expression. IRCAM Centre Pompidou, Paris, France, pp 208–211

Jorda S, Geiger G, Alonso M, Kaltenbrunne M (2007) The reactable exploring the synergy between live music performance and tabletop tangible interfaces. In: the Proceedings of the 1st Conference on Tangible and Embedded Interaction, New York, NY, USA

Kerckhove D (1993) Touch versus vision: Ästhetik neuer technologien. Die Aktualitat des Ästhetischen,Germany, pp 137–168

Le Groux S, Manzolli J, Verschure P (2010) Disembodied and collaborative musical interaction in the multimodal brain orchestra. In: Proceedings of the International Conference on New Interfaces for Musical Expression, pp 309–314

Lee JS, Tatar D, Harrison S (2012) Micro-coordination: because we did not already learn everything we need to know about working with others in kindergarten. In Proceedings of the ACM 2012 conference on Computer Supported Cooperative Work, CSCW '12, New York, NY, USA, pp 1135–1144

Lerdahl F, Jackendoff R (1996) A generative theory of tonal music. The MIT Press, Cambridge

Leslie G, Mullen T (2011) MoodMixer: EEG-based collaborative sonification. Proceedings of the International Conference on New Interfaces for Musical Expression, 30 May–1 June 2011. Oslo, Norway, pp 296–299.

Levin G (2006) The table is the score: an augmented-reality interface for real-time, tangible, spectrographic performance. In: Proceedings of the International Computer Music Conference (ICMC' 06), New Orleans, USA

Maxwell J, Pasquier P, Eigenfeldt A (2009) Hierarchical sequential memory for music: a cognitive model. In: the Proceedings of the 10th International Society for Music Information Retrieval Conference. ISMIR 2009, Kobe International Conference Center, Kobe, pp 429–434

Maxwell JB, Eigenfeldt A (2008) The MusicDB: a music database query system for recombinance-based composition in Max/MSP. In: the Proceedings of the International Computer Music Conference, ICMC2008. (Published Online)

McDermot J, Oxenham AJ (2008) Music perception, pitch and the auditory system. In: the Proceedings of the Current Opinion in Neurobiology 18(4):452–463 (geen locatie!)

Miller K (2009) schizophonic performance: guitar hero, rock band, and virtual virtuosity. J Soc Am Music 3(4):395–429

Miranda E (2006) Brain-computer music interface for composition and performance. Int J Disabil Hum Develop 5(2):119–126

MIROR website (2012) Musical interaction relying on reflection. http://www.mirorproject.eu

Munoz EE (2007 Nov) When gesture sounds: bodily significance in musical performance. In Conference proceedings from the International Symposium on Performance Science, ISPS2007. Porto, Portugal, pp 55–60

Nijholt A, Tan DS, Allison BZ, Millán José del R, Graimann B (2008) Brain-computer interfaces for HCI and games. CHI Extended Abstracts 2008:3925–3928

Pachet F (2002 Sep) The continuator: musical interaction with style. In: Proceedings of the International Computer Music Conference. ICMA2002. Gothenburg, Sweden, pp 211–218

Paine G (2009 Aug) Towards unified design guidelines for new interfaces for musical expression. Organ Sound 14(2):142–155. (Published Online)

Patten J, Recht B, Ishii H (2002) Audiopad: a tag-based interface for musical performance. In: Proceedings of the 2002 conference on New interfaces for musical expression. National University of Singapore, Singapore, pp 1–6

Poupyrev I, Berry R, Kurumisawa J, Nakao K, Billinghurst M, Airola C, Kato H, Yonezawa T, Baldwin L (2000) Augmented groove: collaborative jamming in augmented reality. In: ACM SIGGRAPH2000 Conference Abstracts and Applications, New Orleans, USA, p 77

Raffa R (2011) Rhythmsynthesis: visual music instrument. In: Proceedings of the 8th ACM conference on Creativity and cognition. ACM, pp 415–416

Reidsma D, Nijholt A, Bos P (2008 Dec) Temporal interaction between an artificial orchestra conductor and human musicians. Computers in Entertainment 6(4):1–22

Rosenbaum E (2008) Melodymorph: a reconfigurable musical instrument. In: Proceedings of the 8th International Conference on New Interfaces for Musical Expression. NIME2008. Genova, Italy, pp 445–447

Sänger J, Müller V, Lindenberger U (2012) Intra- and interbrain synchronization and network properties when playing guitar in duets. Front Hum Neurosci. 2012; 6: 312. Published online 2012 November 29. Prepublished online 2012 July 6. doi: 10.3389/fnhum.2012.00312

Takahashi M, Kumon Y, Takeda S, Inami M (2011) Remote hand clapping transmission using hand clapping machines on live video streaming. In: Proceedings of Entertainment Computing

Tarumi H, Akazawa K, Ono M, Kagawa E, Hayashi T, Yaegashi R (2012) Awareness support for remote music performance. Advances in Computer Entertainment, ACE2012, Nepal, pp 573–576

Tatar D (2012) Human-computer interaction seminar. University Lecture, Stanford

Teitelbaum R (1976) In tune: some early experiments in biofeedback music (1966–1974). Vancouver, BC, Canada: Aesthetic Research Center of Canada Publications

Thompson WF, Russo FA, Quinto L (2008) Audio-visual integration of emotional cues in song. Cogn Emot 22(8):1457–1470, Published Online

Varni G, Camurri A, Coletta P, Volpe G (2008) Emotional entrainment in music performance. In: Proceedings of the 8th IEEE International Conference on Automatic Face & Gesture Recognition. FG2008. Amsterdam, The Netherlands, pp 1–5

Varni G, Volpe G, Camurri A (2010) A system for real-time multimodal analysis of nonverbal affective social interaction in user-centric media. IEEE Transactions of Multimedia, vol 12, pp 576–590. (Published Online)

Verbeek P-P (2005) What things do: philosophical reflections on technology, agency and design. Science and Engineering Ethics, Pennsylvania State University Press

Verplank Bill (2011) Keynote on motors and music. Presented on Chi Spark 2011 as the second keynote

Verplank Bill, Gurevich M, Mathews M (2002) The plank: designing a simple haptic controller. In: Proceedings of the 2002 conference on New interfaces for musical expression. NIME2002. Singapore, pp 1–4

Ward KPN, O'Modhrain S (2008) A study of two thereminists: towards movement informed instrument design. In: Proceedings of the 2008 Conference on New Interfaces for Musical Expression. NIME '08. Genova, Italy, pp 117–121

Weinberg G, Driscoll S (2006) Robot-human interaction with an anthropomorphic percussionist. In: Proceedings of the SIGCHI conference on Human Factors in computing systems. CHI2006. New York, USA, pp 1229–1232

Chapter 5
Improvising with Digital Auto-Scaffolding: How Mimi Changes and Enhances the Creative Process

Isaac Schankler, Elaine Chew and Alexandre R. J. François

5.1 Introduction

This chapter examines the creative process when a human improviser or operator works in tandem with a machine improviser to create music. The discussions are situated in the context of musicians' interactions with François' Multimodal Interaction for Musical Improvisation, also known as Mimi (François et al. 2007; François 2009).

We consider the questions: What happens when machine intelligence assists, influences, and constrains the creative process? In the typical improvised performance, the materials used to create that performance are thought of as tools in service of the

---

This chapter incorporates, in part and in modified form, material that has previously appeared in "Mimi4x: An Interactive Audio-visual Installation for High-level Structural Improvisation" (Alexandre R. J. François, Isaac Schankler and Elaine Chew, International Journal of Arts and Technology, vol. 6, no. 2, 2013), "Performer Centered Visual Feedback for Human-Machine Improvisation" (Alexandre R. J. François, Elaine Chew and Dennis Thurmond, ACM Computers in Entertainment, vol. 9, no. 3, November 2011, 13 pages), "Preparing for the Unpredictable: Identifying Successful Performance Strategies in Human-Machine Improvisation" (Isaac Schankler, Alexandre R. J. François and Elaine Chew, Proceedings of the International Symposium on Performance Science, Toronto, Canada, 24–27 August 2011), and "Emergent Formal Structures of Factor Oracle-Driven Musical Improvisations" (Isaac Schankler, Jordan L.B. Smith, Alexandre R. J. François and Elaine Chew, Proceedings of the International Conference on Mathematics and Computation in Music, Paris, France, 15–17 June 2011).

I. Schankler (✉)
Process Pool Music, Los Angeles, CA, USA
e-mail: eyesack@gmail.com

E. Chew
Queen Mary University of London, London, UK
e-mail: elaine.chew@qmul.ac.uk

A. R. J. François
Interactions Intelligence, London, UK
e-mail: alexandrefrancois@gmail.com

N. Lee (ed.), *Digital Da Vinci*, DOI 10.1007/978-1-4939-0536-2_5,
© Springer Science+Business Media New York 2014

performer's will. When a machine is a central partner in music creation, this division of labor is not so clear-cut. In this scenario, can the performer truly be said to be in control of the creative output? This suggests a model of creativity in which ideas do not necessarily originate from the human operator, but instead emerge from the interaction the operator has with another entity or environment. Mimi is an ideal subject for exploring this interaction because it sometimes seems to function as a tool, and at other times as an operator.

In the first part of this chapter, we describe Mimi in the context of other human-machine improvisation systems, outline some strategies for successfully perform-ing with Mimi, and discuss the formal structures that emerge from the interactions between Mimi and a human performer. In the second part of this chapter, we de-scribe Mimi4x in the context of models for group improvisation, and discuss how Mimi4x facilitates the redistribution of musical creativity between a composer, a user, and the system itself.

5.2   The Mimi System

Mimi was designed to enable human-computer musical improvisations in such a way as to reflect and interrogate the improvisational practice of the performer. Em-ploying the principle of recombination, Mimi takes musical input from a performer and dynamically re-organizes the material to generate novel musical output in the midst of a performance.

One goal of the Mimi project is to explore different modes for communication in interactive music systems and different representations for the musical creation process. In the right hands, Mimi becomes an invaluable tool for examining the mu-sician's own improvisational habits. Distinctive aspects of Mimi's behavior can also influence and even dictate the actions of the performer, raising questions about cre-ative ownership. A number of human-machine improvisation systems exist, many of which will be described in Sect. 2.1, Related Work on Improvisation Systems. Distinct from most other systems, Mimi allows the performer to track the current state of the system, consisting of a 10-s preview of the future and a 10-s review of the past. The visual interface acts as a window into the machine's "mind."

The Mimi system has evolved over the years, and even extended to include mul-tiple instantiations of the original Mimi. Mimi 1.0 was the result of the playful interactions between computer scientist/amateur musician (François), an engineer-musician (Chew), and a piano pedagogue (Dennis Thurmond). Mimi's design was further refined in collaboration with a composer/improviser (Schankler), resulting in Mimi 1.5 (shown in concert in Fig. 5.1). This collaboration also resulted in Mim-i4x, an offshoot of Mimi that allows users without musical training to interact with multiple Mimi improvisational streams.

Mimi was created using the formalism defined by François' Software Architec-ture for Immersipresence (SAI) framework as described in (François et al. 2011). SAI's architectural style (François 2004) allows for the integrated processing of

Fig. 5.1 Mimi 1.5 in performance with Schankler, at Boston Court Performing Arts Center in Pasadena, California, on June 5th, 2010

persistent data, and volatile information streams, thus enabling the concurrent manipulation of persistent symbolic data structures in Mimi, specification of an interactive and synchronous visual display, and synthesis of "improvised" MIDI output.

5.2.1 Related Work on Improvisation Systems

This section relates Mimi to representative past and ongoing work in interactive machine improvisation, focusing on the different interaction modalities.

Voyager: George Lewis has been creating and performing with algorithmic improvisation systems since the 1980s. One example is Voyager, which Lewis describes as a "nonhierarchical, interactive musical environment that privileges improvisation" (Lewis 2000). In Lewis' approach, the preparation process consists of creating of a computer program. During the performance, the program automatically generates responses to the musician's playing, as well as new material. The performer listens and reacts to the system's musical output, closing the exclusively aural feedback loop. The musician has no direct control over the program during performance, i.e. the interaction between the performer and Voyager is purely sonic.

Since it does not strictly require performer input, Lewis views performing with Voyager as enacting a "subject-subject" relationship instead of a "stimulus/response" one. Superficially, since Mimi uses the performer's musical material as the foundation of its improvisation, a performer enters into a stimulus/response relationship with Mimi. In a sense, Mimi acts as a distorted mirror by reflecting back the actions of the player in a recombined order. On a deeper level, however, the way in which Mimi reflects and recombines determines a great deal of how the musical content is delivered and perceived, as we will see. Mimi's behavior, in other words, is just as important, if not more important, than the material the player provides.

OMax: Mimi's improvisation engine is inspired by that of the OMax family of interactive improvisation systems (Assayag et al. 2006). In OMax, Assayag, Dubnov et al. introduce the factor oracle approach in the context of machine improvisation (Assayag and Dubnov 2004; Dubnov and Assayag 2005). Evolutions of OMax added capability to directly learn from and generate an acoustic signal (Assayag et al. 2006). Most improvisation systems, including the version of Mimi described here, deal with symbolic (MIDI or MIDI-like) representations for input and output. Recent versions of OMax can also learn and generate music from a timbral descriptor-based model in addition to note-level representations.

OMax uses a combination of OpenMusic (Assayag et al. 1999; Bresson et al. 2005) and Max/MSP (www.cycling74.com) to handle static data and real-time information, with OpenSound Control (opensoundcontrol.org) serving as the communication and coordination protocol. In OMax, offline versions of the learning and generation algorithms are implemented in OpenMusic. OpenMusic's functional programming approach does not provide the means to specify the temporal behavior of a system, a requirement for the correct and efficient handling of real-time (or online) events. Consequently, OMax relies on Max/MSP to handle online aspects such as real-time control, MIDI and audio acquisition, and rendering. OpenMusic and Max/MSP adopt similar visual metaphors (patches), but with different and incompatible semantics. As observed by Puckette (2004), Max patches contain dynamic process information, while OpenMusic patches contain static data. Therefore, communication between and coordination of the two subsystems in OMax requires the use of the interaction protocol, OpenSound Control.

In contrast, Mimi is designed and implemented in a single formalism (SAI), which results in a simpler and more scalable system. For example, complex visualization functionalities can be readily integrated in an efficient and seamless way, without compromising system interactivity. A recent addition in the OMax family, WoMax (OMax without Open Music) (Lévy 2009), aims to port OMax's main OpenMusic components to Max/MSP in order to facilitate the visualization of the oracle structure and of the traversal process during improvisation.

In all the interactive improvisation systems that we have surveyed, the improvising musician interacts with the system based on purely aural feedback. In OMax, further dynamic capabilities are accorded the computer through a human operator who controls the improvisation region in the oracle, instrumentation, etc., through a visual interface during performance. This control mechanism is necessary because the system learns continually from the performer, as well as the music streams it generates, leading to a proliferation of musical ideas that could create a challenging improvisation environment if left unconstrained. WoMax introduces a visualization component that aims to facilitate the operator's understanding of the oracle's structure and to improve interactive control of the system.

Mimi explores a different approach, with interaction specifically designed for and under the sole control of the improviser. In the designs of existing systems, the performer relies almost exclusively on real-time auditory feedback to assess the state of the system and to gauge its possible evolution in the near future. In Mimi, visual feedback of future and past musical material and of high-level structural information provides timely cues for planning the improvisation during performance. When the oracle on which the improvisation is based is laid down in the preparatory stage, it can be spontaneously generated or pre-meditated and carefully planned.

Factor oracle-based improvisation is based on a stochastic process in which the musical material generated is a recombination of musical material previously learned. (Note that the learning of the repetition structure is deterministic.).

Other: Many improvisation systems first make use of various probabilistic models to learn parameters from large bodies of musical material, then generate new material with similar statistical properties, akin to style. Examples of machine-improvisation systems employing probabilistic models include Pachet's Continuator (Pachet 2003), Thom's Band-OUT-of-the-Box (BoB) (Thom 2000a, 2003), and Walker's ImprovisationBuilder (Walker 1997; Walker and Belet 1996; Walker et al. 1992). Musical interaction modalities during performance vary from system to system and range from turn-taking dialogue to synchronous accompaniment, but, in all cases, performer-machine interaction is based exclusively on aural feedback.

Weinberg and Driscoll (2006) introduced Haile/Pow, a humanoid-shaped robotic drummer that listens to the performer, learns in real time, and plays along in a collaborative fashion. In this scenario, the performer may be able to anticipate, to a small extent, the robot's next gesture by closely watching the robot, but the primary feedback remains auditory. The difference here is that the robot is the follower, and the human musician the leader.

In the use of Mimi considered here, the musician is first the leader in the sense that s/he lays down the seed musical material on which the machine's improvisation is based. Once the machine begins extemporizing, it then becomes the leader, producing the music streams over which the musician will then improvise. This is true to some extent for the Voyager and OMax systems as well.

5.2.2   Mimi's Visual Interface

The key and novel component of the Mimi system is its visual interface. The screen image is divided into two halves. The upper half, labeled Performance visualization, displays a scrolling piano roll notation. The central red line represents the present moment, with notes to come on the right and notes recently sounded on the left. This upper display acts as a map that allows the performer to plan ahead and to revisit recent past material. In Mimi 1.0, the machine-generated improvisation is colored blue, while the human improviser's notes are colored red (see Fig. 5.2).

Mimi 1.5 uses Russian composer Alexander Scriabin's color-to-tone mapping (Galeyev and Vanechkina 2001). Scriabin's system is based on the circle of fifths, making it ideal for performance visualization, allowing the performer to glean harmonic information intuitively and at a glance (see Fig. 5.3). For example, white keys tend to be bright colors, and black keys tend to be pastels.

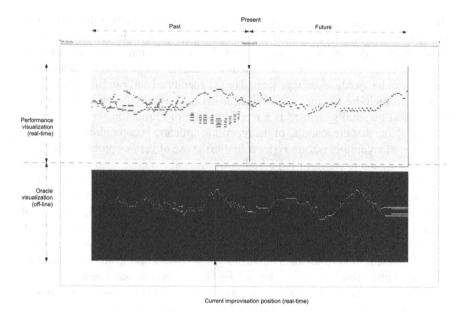

Fig. 5.2 Visual interface of Mimi 1.0

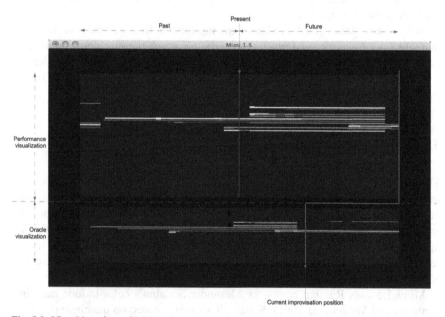

Fig. 5.3 Visual interface of Mimi 1.5

The lower half of the screen, labeled Oracle visualization, shows the preparatory material that the system recombines to generate new music. A red cursor shows the machine improviser's present position in the preparatory material. This lower display presents a visual aid to the human improviser, helping the improviser keep a mental note of the structural content and relative location (within the initial material) of the current machine improvisation. When projected on a large screen in a performance, these visuals may also assist the audience in perceiving higher-level structures and the subtleties of improvisation strategies.

5.2.3   Interacting with Mimi

In the context of the original experiments with Mimi 1.0 and Dennis Thurmond, interactions with the system occurred in two distinct phases: preparation and performance. Interactions in the two stages occur over identical modalities, but under different external constraints.

In the preparation phase, a musician lays down the material that will be used by the system for generating improvisations. This material from the human performer/improviser may result from a spontaneous creation, or it may result from a carefully planned process, more akin to composition. As the initial material is created, Mimi's visual interface shows the notes played by the performer in piano-roll representation, scrolling from right to left in the upper screen and collecting cumulatively in the lower screen.

In the performance phase, the system generates new musical material from the prepared material (visible on the lower screen), with which the performer improvises. In addition to providing auditory feedback, the upper half of the visual display shows the interactions between the machine's improvisations and the human performer's improvisations, as well as the musical material to come and recently passed in real time. The lower half of the visual display documents the current state of the improvisation engine with respect to the preparatory material.

In Mimi 1.5, the preparation and performance phases are concurrent and may overlap, i.e. Mimi may be learning new material and improvising at the same time. A MIDI controller provides an intuitive interface for the performer to have quick access to commands and parameters before and during performance. The performer may trigger Mimi's learning state or improvisation state on or off at any time. There are also controls to clear Mimi's memory, to adjust the recombination rate, and attenuate the playback volume. This interface has powerful implications for performance strategies and formal structures that will be discussed in the next two sections.

5.3 Performance Strategies

Improvisations performed with human-machine systems like Mimi present their own distinct set of challenges and benefits for the human performer apart from the issues that typically arise in improvisation between human performers. For example, the window into the future that Mimi displays can provide a visual aid ordinarily unavailable to human performers, and yet Mimi cannot react spontaneously to unfolding circumstances in the moment as a human performer would. In the following section, we outline some strategies that can help guide a performer in both preparing for, and playing with, a system like Mimi. Both performers (Thurmond and Schankler) who had significant experience practicing with the system employed similar strategies that diverged in a few significant ways.

5.3.1 Performing with Mimi 1.0 (Dennis Thurmond)

The balance of musical ideas in the oracle is important to the success of an oracle. Preparatory material that is too dense, either vertically or horizontally, will lead to machine improvisations that leave the performer no room to get a note in edgewise. One sure way to make the interaction fail is to play an existing dense and complete composition. On the other hand, leaving too many holes in the preparatory material can also be detrimental to human-machine improvisation. The density of the preparatory material is readily assessed through the visual interface.

Three design sessions with Thurmond were recorded in January and February 2007 (viewable online at www.youtube.com/user/alexandrefrancois). In each session, Thurmond first performed with the system, and then answered questions about the experience from Chew. In the first documented design session, Thurmond speaks on the importance of incomplete structures and open spaces in the design of preparatory material (emphasis added):

> [T]he first thing I did was I made a structure that was only partially complete. In other words I intentionally used a certain kind of formula, compositional formula, of certain dissonances combined together. Then, I would leave open spaces, and then I would try to change the registers using that same melodic model.

(Unless otherwise noted, the quotes in the remaining parts of this section are the words of Dennis Thurmond.)

5.3.2 Anticipating the Future

When human musicians improvise together, they are able to anticipate what the other person is about to do and to accommodate it in their response. Furthermore, the musicians have agreed upon the structure of the piece and have a general map for what is going to happen. Mimi's visual interface, both the future half of the

Performance visualization panel and the Oracle visualization panel, helps create an environment that fosters this kind of anticipatory behavior. Being able to foresee the machine's near future improvisations is like having a window into one's improvisation partner's head.

> [I]f you're improvising with another improviser it's like you're able to look into their mind ahead of time and see what they are about to do.

With this information about the future, the human improviser can then anticipate the machine's improvisations so as to plan counter material in response.

> [I]f the oracle decides that it wants to extend something, or put something out a little bit differently, I see it coming and I can prepare for that.

Another aspect of ensemble performance is the visual (gestures) and auditory (breathing) cues that signal the timing of musical entrances. This aspect is not captured in auditory-only feedback interactive improvisation systems. In Mimi, the Future panel allows the human improviser to plan an entrance, thus creating a semblance of the human element of coordination of timing.

> Having the screen is very critical; it makes it two living entities, instead of a random thing, and hoping something comes out of it.

Knowing the system state is an important component of anticipating the future. Thur- mond uses the Oracle visualization panel, in the lower half of the display, to contextualize the present musical content and state of the machine improvisation.

> I wanted to see what [Mimi] was thinking.... it allowed me a lot more freedom than trying to remember just exactly what kind of pattern I set up. So the structure that I set up, I am constantly reminded of the structure.

The oracle's state thus provides the performer with a quick reference for the structure of the present improvisation material.

5.3.3   Reflecting on the Past

We discovered that past information is also important to improvisation; something that should not have come as a surprise in hindsight. An understanding of history, albeit a short one, allows the improviser to solve future problems. The visual interface makes it easy to introduce the reflective component into human-machine improvisation.

> [S]ometimes when you are improvising, you're not sure what it is exactly that you're doing. But with the history screen, I can see what I have just done and I can use that, and build on that.

One reason to see what has transpired is to be able to repeat something that is not quite right so that it turns into a feature.

[H]aving a history is very important because if I do something that is not quite right, in improvisation if you make a mistake, then you don't jump away from it, you stay with it to make it not a mistake any more.

The statement resonates with John Cage's quote, "The idea of mistake is beside the point, for once anything happens it authentically is." Another reason to keep track of recent history is to be able to review the interaction in order to repeat an interaction that was particularly successful.

Because if you have done something by accident that's really good, then you want to see how it lines up, how it comes back again, and you want to use it again, and maybe elaborate on it a little bit.

When the unexpected happens, to recreate it again, one has to recognize how it happened and in what context, so as to be able to repeat a particular pattern in the future. Comparing the importance of the three types of information—future and past material and oracle state—Thurmond estimates that he pays attention to the future content part of the panel approximately 60 % of the time, and divides the remaining 40 % of the time between the past and the oracle state parts of the display. He admits that these proportions can be subject to change with experience with Mimi and for a particular oracle:

I think that the weighting of the screens will change as the person becomes more and more familiar with it, and it becomes a lot more common to the performer. Then, I think the [ratio of time spent on the] two screens will equal out. This bottom screen will turn out to be less important as the same oracle happens over and over. You just use it as a quick reference, so it goes down to just a few percent after repeated performance.

5.3.4   Performing with Mimi 1.5 (Isaac Schankler)

In the subsequent version of Mimi devised in collaboration with Schankler, the color-to-tone mapping, concurrent preparation and performance, and added volume and recombination rate parameter controls opened up new possibilities and strategies for improvisation. Most apparently, the volume parameter allowed for the possibility of less abrupt transitions (i.e. fading in and out), while the recombination rate parameter could be used to influence how often Mimi would make jumps within its data structure. Informally, the recombination rate can be thought of as a parameter affecting degrees of order and chaos.

New techniques and strategies, described in the following section, were discovered over the course of several sessions practicing and performing with the system. Examples in the this section are taken from live performances at the People Inside Electronics concert at Boston Court Performing Arts Center in Pasadena, CA on June 5, 2010 (viewable at vimeo.com/15379437), and the Ussachevsky Memorial Festival concert at Pomona College on February 4, 2011 (viewable at vimeo.com/mucoaco/mimi1).

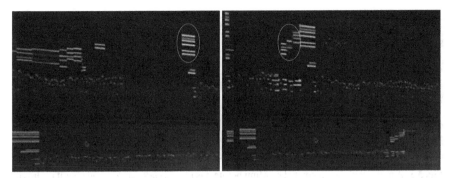

Fig. 5.4 An abrupt chord entrance by Mimi before it is played (*left, circled*), and after it is played (*right*), with connecting chords by a human performer (*circled*)

5.3.5   Discontinuities and Interpolations

As Thurmond states, Mimi's visualization scheme affords the performer unique musical opportunities not ordinarily available in the course of improvisation. Improvising musicians are not normally privy to the exact details of the immediate plans of other improvisers, but with Mimi, the improviser is also able to plan a precise reaction to musical material before it sounds, adding a novel element to the interactions.

Mimi may occasionally recombine two segments of material that seem disjunct in terms of range or dynamics. The discontinuities happen because in Mimi's factor oracle model, octave transpositions are semantically identical. In other words, Mimi does not differentiate between a C at the bottom of the piano and a C in the extreme upper register. In terms of harmony and tonality, this is a sensible decision, since the same tone in a different octave is functionally equivalent. However, in the context of Mimi's performance, this will sometimes have the effect of creating unusual and unexpected leaps between different registers in the middle of phrases.

Confronted with this, the performer has several different options for how to react. The performer may simply ignore the discontinuities, regarding them as mistakes. However, continuing to interpret them in this way has the potential for extreme frustration in the context of an ongoing improvisation. Since these discontinuities are an intrinsic aspect of Mimi's "sound," interpreting them as errors would not help a performer successfully negotiate an improvisation with Mimi.

Another option is for the performer to "smooth out" the boundary through the interpolation of connecting material. Figure 5.4 shows one instance of such a boundary, where the performer plays a series of ascending chords to lead up to what would have otherwise been an abrupt chord onset.

This kind of interpolation does not necessarily have to be limited to instances where discontinuities are glaringly obvious. The performer may choose to interpolate material that leads up to any musical event, or departs from an event, or links two events. Figure 5.5 shows one example of a musical phrase, generated by Mimi, which is then elaborated on with interpolations improvised by a human performer

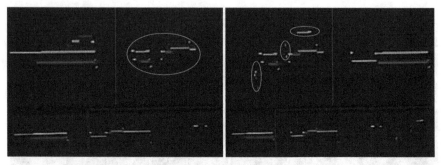

Fig. 5.5 A musical phrase generated by Mimi before it is played (*left, circled*) and after it is played (*right*), with interpolated material by a human performer (*circled*)

both before and during the phrase. The final result is a synthesis of two or more musical ideas that nonetheless can be perceived as a single phrase.

If executed proficiently, the resulting performance differs from a typical duet, in that it may be difficult or even impossible for a listener to distinguish between the performer's actions and Mimi's. It is neither a "subject-subject" relationship or a "stimulus/response" one, but a more ambiguous third category where human and computer counterparts work together to create a chimeric whole. Thus, the performer's actions can become intertwined with Mimi's in an intimate and almost seamless way.

Yet another option is to incorporate leaps and jumps into one's own playing, and choose musical material that are more conducive to this kind of movement. In this scenario, Mimi dictates a distinctive stylistic feature of the collective performance, with the performer following Mimi's lead.

5.3.6   Transitions

Transitions between different musical sections or ideas present particular challenges for the performer improvising with Mimi, since the performer has no direct control over what Mimi chooses to play, and Mimi does not respond directly to the performer as a human improviser would. For example, a human improviser would be likely to respond immediately to a change in harmony or texture, while Mimi cannot do so (though Mimi may learn from that material and perform that change at some point in the future). Therefore the performer who wishes to incorporate musical transitions into an improvisation must use more indirect methods.

It is possible to create gradual harmonic transitions through the use of common tones or common chords. For example, the performer can at any time introduce new tones that complement the existing texture. As Mimi incorporates these new tones into its data structure, over time it may move to introduce these new tones into its musical tapestry, and the performer may move to a different sonority that incorporates those tones. By stringing several transitions of this sort together, long-term

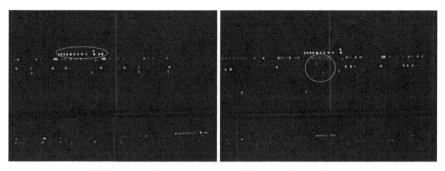

Fig. 5.6 A new pitch introduced by the human performer (*left, circled*) and then reintroduced by Mimi (*right*), with a new harmonization by the human performer (*circled*)

harmonic progressions can be created, if the performer so wishes. Figure 5.6 demonstrates one example of a common-tone transition.

Mimi's controls also give the performer the option of clearing Mimi's memory of all musical material. This is a more drastic transition strategy, but it is effective in creating large structural boundaries in an ongoing improvisation.

5.4   Formal Structures

Issues relating to formal design of improvisations with Mimi are worth discussing in greater depth. Figure 5.7 shows an analysis of a performance with Schankler and Mimi at the People Inside Electronics Inaugural Concert at Boston Court Performing Arts Center in Pasadena, CA on June 5th, 2010. This analysis includes Schankler's informal perceptions of what occurred during the performance, as well as descriptions of what musical materials Mimi and Schankler were playing.

The following section describes various formal structures observed in this and other improvisations with Mimi. The fact that large-scale forms can be identified in improvisations with both Mimi and a human performer gives rise to a question: do these forms arise from the human performer's conscious or unconscious intent, Mimi's behavior, or some combination of the above?

In this case, the human performer had improvised extensively with Mimi over several months before performing the improvisations analyzed here. During this time, the performer adapted his usual improvisational technique in order to create successful performances with Mimi (since Mimi is incapable of certain kinds of adaptation, like rhythmic or harmonic coordination with the human performer). In some cases, these adaptations even ran counter to the performer's usual technique (for example, certain kinds of harmonic movement would be restricted by Mimi's behavior). In other words, while the performer had significant control over the improvisation's moment-to-moment content, Mimi had much more influence over the structure of that content.

Time into piece		Isaac's Notes	Mimi				Isaac
Minutes	Seconds		Learning	Improvising	Memory	Playing	Playing
0	0	I start playing and Mimi starts learning.					voice part a
0	10	Mimi starts improvising (10-second delay).				voice part a	voice part b
0	22	Mimi starts playing. I play a lower voice in counterpoint.				voice part b	voice part c
0	37	Mimi starts playing the lower voice, so I switch to playing another upper voice.				voice part c	octave bass
0	53	Mimi switches to the upper voice, so I start playing bass tones with octaves.				voice a/b	octave treble
1	5	Mimi switches to the first voice, and I start playing an upper voice in octaves.					
1	17	Mimi stops learning.					ornamented idea
1	27	Mimi starts learning again, and I play a new, more ornamented idea.				ornamented idea	counterpoint
1	55	Mimi starts playing the new idea, and I play a lower counterpoint to it.				ornamented idea	ornament + wipe out
2	0	As the music builds, Mimi finishes immediately, and her memory is cleared.			x		chordal idea
2	5	I play a new, chordal idea.					
2	10	Mimi starts learning and improvising.				chordal idea	
2	36	Mimi stops learning.					
2	50	Mimi's memory is cleared and she starts learning the new, more active idea I'm playing.					active bass
3	5	I see that Mimi's about to play the big chord, and try to play a lead-up to it.			x	big chord	lead up to big chord
3	10	Mimi picks up on the active bass idea, and I play high pointy chords as counterpoint.				active bass	pointy chords
3	34	Mimi finishes immediately and I play a chordal interlude.					chordal interlude
3	39	Mimi stops learning and starts improvising again.					
3	50	Mimi goes right back into the active bass idea. I play pointy chords again.				active bass	pointy chords
4	17	Mimi's memory is cleared.				active bass	4-note figure
4	20	I start playing a four-note figure and hold down the pedal to transition into a new section.					
4	30	Mimi starts learning.					
4	46	Ouch! A sharp attack from Mimi before it starts playing the four-note figure.				4-note figure	4-note figure
4	50	I play above and below Mimi's four-note figure.				4-note figure	above-and-below
5	13	Mimi stops learning.				4-note figure +	above-and-below
5	17	Mimi starts learning. I play a new rhythmic bass idea.				4-note figure +	rhythmic bass
5	45	I slow down to match Mimi's melodic descent.				4-note figure +	slow down
5	50	Mimi stops learning.					
5	54	Mimi picks up the rhythmic bass idea. I play a pick-up to transition into a noodly melody against Mimi's bass.				rhythmic bass	noodly melody
6	12	Mimi starts learning. I play a new version of the rhythmic idea in the treble range.				rhythmic bass	rhythmic treble
6	30	I play upward "swoop" gestures, and a few other ideas in quick succession.				rhythmic bass	upward swoops +
6	42					rhythmic bass	pointy chords
6	47					rhythmic bass	new motif repeated
6	59	Mimi plays the "swoops," and I play a low countermelody.				upward swoops	low noodly melody
7	6	Mimi and I both play pointy chords.				pointy chords	pointy chords
7	11	Mimi returns to the bass idea, and I return to melody.				rhythmic bass	noodly melody
7	24	I sit out while Mimi plays the "swoops" again.				upward swoops	
7	30					pointy chords	
7	38	Mimi plays the slowing down idea from earlier, finishes immediately and its memory is cleared.			x	slow down	
7	50	Mimi starts learning.					
8	0	I play rising gestures and Mimi starts improvising.					pedaled notes
8	8	Mimi starts playing the rising gestures.				rising gestures	rising gestures
8	20	Here I'm just playing more and more dramatic gestures, trying to get a big climactic sound.				rising gestures	more rising gestures
8	30	A few of Mimi's notes "hang over" the last chord, so I play a few notes in response to complete the idea.				rising gestures	more dramatic gestures
8	40	It's over! Except Mimi's still learning…					few notes to finish idea

Fig. 5.7 Analysis of performance with Schankler and Mimi. An open bracket indicates a process starting, while a closed parenthesis indicates a process ending. An "x" indicates Mimi's memory being cleared

Performer:	A	B	C	D	E	F	etc.
Mimi:		A	B	C	D	E	etc.

Fig. 5.8 Canon with Mimi and human performer

Performer:	A	B	C	D	E	F	etc.
Mimi:		A	B	C	A	D	etc.

Fig. 5.9 Canon-like form with Mimi and human performer

For the remainder of this section, we will discuss some of the formal structures observed in these improvisations with the assumption that they emerge, at least in part, from Mimi's inherent behavior. However, whether this is a truly intrinsic aspect of Mimi's behavior or a complex interaction between Mimi and the personal tendencies of the performer is a definite area for further study with more performers and performances.

5.4.1 Imitative Forms

The canon, a polyphonic texture in which a second voice exactly imitates the first, is fundamentally connected to Mimi's performance habits. By default, Mimi's actions are delayed by 10 seconds to give the performer a visual heads-up of what Mimi is about to perform. If Mimi chooses not to recombine any of the subsequent material (a distinct possibility, especially if the recombination rate is low), Mimi's performance will be an exact copy of the original performance, delayed by 10 seconds. In other words, Mimi will create a canon at the 10-second level with the original performer, regardless of what the performer chooses to play (see Fig. 5.8).

This type of figure can also be seen in the first minute of the performance analyzed in Fig. 5.7. Even when this exact sequence does not occur, Mimi's habits create an abundance of canon-like, or heavily imitative, forms. Imitative forms can be described as those in which fragments of musical material are passed from one voice to another to another in a polyphonic texture (in this case, passed from the original performer to Mimi) (see Fig. 5.9).

5.4.2 Rondo-like Forms

As Mimi begins to revisit different sections of musical material, more formal resemblances may emerge. In its conception, Mimi does not privilege any material over any other, but Mimi's design allows it to learn continuously from the performer during the course of an improvisation (if the performer chooses to allow it). In this

Performer:	A	B	C	D	E	F	G	etc.
Mimi:		A	B	A	B	A	C	etc.

Fig. 5.10 Rondo-like form with Mimi and human performer (through-composed)

Performer:	A	B	C	A	D	A	E	etc.
Mimi:		A	B	A	C	A	D	etc.

Fig. 5.11 Rondo-like form with Mimi and human performer (rondo-like)

Performer:	A	B	C	A	D	E	F	G	E	H	etc.
Mimi:		A	B	A	C	E	F	E	G	etc.

Fig. 5.12 Large-scale binary form with nested rondo-like forms

scenario, material that enters into Mimi's memory earlier will be more likely to be heard by virtue of simply being in memory for a longer period of time. A is more likely to be heard than B, which is more likely to be heard than C, and so on. As a result, the very first thing Mimi hears is likely to be the most significant piece of musical material. If Mimi revisits this initial idea often enough, it may take on the quality of a refrain, creating a rondo-like form in which the initial idea recurs several times over the course of a performance, interspersed with different, successive episodes (see Fig. 5.10).

This will happen quite often even if the human performer proceeds in a through-composed fashion (not revisiting ideas). However, a sensitive performer may also choose to respond to Mimi's behavior in a way that brings out these formal divisions (see Fig. 5.11).

The performer may also choose to clear Mimi's memory at some point, creating a blank slate to be populated with new material. In this way, the performer is able to create large-scale formal divisions between sections that do not share the same pool of material (see Fig. 5.12).

The prevalence of these structures demonstrates that, while the performer wields much power in shaping a performance with Mimi, Mimi also exhibits a great deal of agency and control over the large-scale structures of those performances.

5.4.3 Discussion of small-scale structure

Up until now, much attention has been paid to the large-scale structures that Mimi tends to create, but not much has been said about the small-scale structures they are comprised of. The large-scale section boundaries indicated in the annotations only capture the most broad structures found in the improvisations. A close examination of a short excerpt of Mimi's performance shows that, on a smaller scale, Mimi

Fig. 5.13 Opening phrases of studio performance with annotation of rhythmic cells. Note values were quantized to the nearest 16th note, while allowing for simple triplet divisions.

operates quite differently than one might expect from looking at the large-scale structure.

In Fig. 5.13, Mimi's opening phrases from the MIDI file of a studio performance (viewable online at vimeo.com/mucoaco/mimi3) are transcribed and quantized. Repetitions of the same motive are bracketed and indicated with the same letter. Here we see at work what Olivier Messiaen termed personnages rythmiques, or rhythmic cells, individuated by particular sonorities, that can be expanded or contracted (Boivin 2007). Mimi performs four distinct cells (A through D) before recombining them in various ways. Contractions of a cell are more common (e.g. the abbreviated form of A at the end of m. 5) but expansions (such as the "stutter" in the middle of the version of C that appears in m. 10) are also possible. Mimi's overwhelming preference for its initial idea can also be observed here—motive A recurs a total of 7 times, versus 4 iterations of B, 2 of C and 3 of D before moving on to a new idea (E).

Unlike the canon and the rondo, this particular kind of juxtaposition of rhythmic cells is regarded as a twentieth century innovation, and Mimi's juggling of these particular cells is not wholly unlike the jagged, erratic rhythms found in the "Danse sacrale" from Stravinsky's Le sacre du printemps, for example. These rhythmic innovations informed Messiaen's conception of personnes rythmiques as well as successive generations of composers who sought a new working theory of rhythm (Boulez 1991). However, since Mimi itself has no theory of rhythm, its improvisations lack the rigorous rhythmic structure and symmetries found in "Danse sacrale" and other composed music. One approach to incorporating rhythmic sensitivity into a factor oracle-based system, which involves suppressing factor links between contexts with grossly different average rhythmic densities, was presented by Assay-ag et al. (2006). Examination of exactly how this or other approaches might operate in the context of the Mimi system is warranted, as well as further study of the mecha-nisms by which the observed large-scale structures emerge from these smaller units.

5.5   Mimi4x and User-Guided Improvisations

Mimi4x is an interactive audio-visual installation based on Mimi 1.5. Mimi was designed with the experienced performer in mind, and the success of its improvi-sations is directly dependent on the skill of the performer. It also presents its own learning curve, and performers benefit greatly from more practice with the system. While this gives musicians opportunities to gain mastery performing with the Mimi system, it also makes Mimi less accessible to non-musicians or even musicians with little improvisational experience.

Mimi4x makes Mimi's inherent improvisational capabilities accessible and ap-parent to all by giving users high-level structural influence over musical material at the structural level, while leaving low-level aspects of musical content up to Mimi. Mimi4x consists of four Mimi instances, each an autonomous improvisation system, seeded with pieces of musical content specifically designed to coordinate with each other. The user controls the musical structure of this content by influenc-ing the activity, volume and recombination rate of each instance using a musical instrument digital interface (MIDI) controller, in this case a Korg nanoKONTROL. Figure 5.14 shows the components of a Mimi4x installation. The user can start and stop each improvised music stream, change its volume (including fade in and out) and manipulate the recombination rate of each improvisation engine. A demonstra-tion video of Mimi4x is accessible online (vimeo.com/mucoaco/mimi4×).

We describe below the two main components of the Mimi4x installation that di-rectly impact the audience and user's experience with the system, namely the visual display and the controls for coordinating and managing the music streams from the four improvisation systems.

5.5.1   Visual Display

Figure 5.15 is a screenshot of Mimi4x showing the four Mimi panels. Each Mimi panel provides a visual interface to an instance of the Mimi 1.5 improvisation sys-tem. Figure 5.15 shows a close up of one Mimi panel. The thickness of each note line segment is proportional to the playback volume of that note. The upper panel demonstrates a fade-out and fade-in effect controlled by the user. A composer/im-proviser prepares original music for each of the Mimi panels (Fig. 5.16).

As with Mimi, the viewer sees 10 s of the music to play and of music recently played. The music displayed in each Mimi panel may be rendered using a different instrument sound setting to differentiate them aurally.

Fig. 5.14 Mimi4x installation components: a monitor or projection screen, visuals from a computer and a MIDI controller

5.5.2 Controls

Figure 5.17 shows the Mimi4x controls on the Korg nanoKONTROL MIDI controller. The user, who may or may not be a musician, uses this controller to coordinate the improvised music streams and to manipulate properties such as recombination rate and volume for each stream.

The Korg nanoKONTROL provides a number of faders, knobs and switches organised in nine identical units. The controls for Mimi4x map to four such units (1–4), each one assigned to one Mimi instance. In each control unit, the lower switch activates and deactivates the improvisation stream of the corresponding Mimi instance. The upper switch stops the improvisation process and clears any music remaining in the future half of the upper pane in the corresponding Mimi panel. The knob controls the recombination rate of the corresponding improvisation engine. The slider controls the volume of the improvised stream and can be used to manipulate the dynamics of the music or fade it out or in.

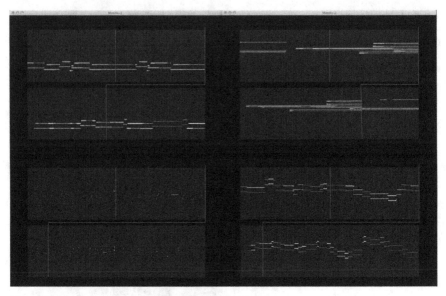

Fig. 5.15 Screenshot of Mimi4x showing four Mimi panels

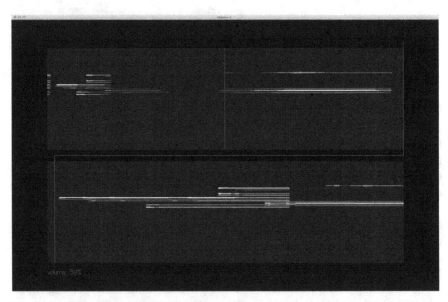

Fig. 5.16 A Mimi window showing system state (*lower pane*), improvised music stream (*upper pane*) and fade-out/fade-in effect

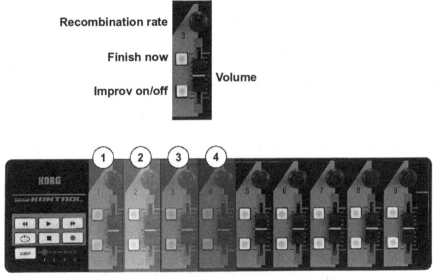

Fig. 5.17 Mapping of Korg nanoKONTROL interface to Mimi4x functions

5.5.3 High-Level Structural Improvisation

Usually, improvisers are guided in performance by some organizing principle like a chord progression. In the absence of such an a priori organizing principle, groups of improvisers must have a common language that allows them to make such organizational decisions on the spot.

Composers such as John Zorn (b.1953) and Walter Thompson (b.1952) have created systems that provide the means for high-level structural improvisation. These systems provide a common language for a guide to make organizational decisions and coordinate improvisers. This guide, in effect, takes on some of the traditional decision-making roles of the composer. Both the guide and the musicians serve as improvisers, but the kinds of decisions made by the guide are different from the decisions made by the instrumentalists. The instrumentalists are generally focused on the moment-to-moment low-level details of music making. The guide has little control over these details, but exercises control over the structural, high-level aspects of the performance, i.e. who should play when and how they should play. In this sense, the guide is also acting as a composer.

In John Zorn's Cobra (Zorn 2004), a prompter cues the players with cards and gestures. Examples include cues for music change (the group of musicians playing should play something different) and group change (the group of musicians playing changes, but the second group has to attempt to play the same music as the first). The instructions define a musical relationship, but not what individual musicians are playing per se. In Walter Thompson's Soundpainting (www.soundpainting.

com), the players are guided by a 'soundpainter' who communicates to the players using a complete system of sign language.

In the Mimi4x installation, the four Mimi instances act as the "musicians," making low-level moment-to-moment decisions about musical materials. The audience member who enters into and participates in the work acts as the guide, making high-level structural decisions by deciding which of the four Mimi instances should be active or inactive and adjusting the recombination rate and loudness levels of all the four. In this way, the user is able to take on some improvisational, compositional and performance roles.

In its modular nature, Mimi4x also bears some resemblance to Terry Riley's influential minimalist composition "In C," in which musicians repeat small cells of music. Each musician progresses through the cells at their own rate, creating an overlapping network of musical layers. A collaborative online music project inspired by Riley's work called "In B Flat 2.0" (Solomon 2009), reveals further resemblances. "In B Flat 2.0" consists of a curated collection of YouTube videos of musicians playing instruments or of background noise, all in the key of B-Flat, arranged in a 5×4 grid. The viewer/user can choose to start or stop any video in the collection, thereby creating a unique texture of different sounds. A volume slider associated with each video panel allows the user to adjust the sound mix.

Giving users influence over mid- to high-level musical structure extends to computer-assisted composition systems. In Hyperscore (www.hyperscore.com, Farbood 2001), the user, through a drawing interface, sketches motives and large-scale structures such as lines indicating when each motif is active, their pitch heights relative to each other and overall perceptions of tension and release. The computer then uses these user-provided guides to generate music. Bamberger (2000) and Hernández's Impromptu (www.tuneblocks.com) allow the user to manipulate the ordering of "tuneblocks" that represent melodic figures and thus exercise control over mid-level structures. As composition tools, neither program was designed for the type of real-time interactive control required in performance.

5.5.4 The Composer's Experience with Mimi4x

Mimi4x presents a novel environment for music making. There are $24(=16)$ possible combinations of improvisation streams at any given time, and even more of the sounded note material and volume levels. This presents new challenges to the composer who prepares the oracles, and the user, a member of the general public who now takes on some of the responsibilities of the composer. This section discusses the implications of this new mode of interaction with music for the composer as well as the user.

Since Mimi4x is not a wholly predictable organism, an effort to write traditional, fully notated music for it would be fruitless. Compositional techniques must be adapted to allow for semi-aleatoric and algorithmically generated events. When creating original music for each Mimi oracle, each oracle in isolation, as well as each of the possible combinations of oracles, must generate compelling and coherent music. This has significant implications for tonal content, rhythm and structure.

Unlike traditional composition, the composer in this case does not write a complete piece, but she/he can lay down templates that may be manipulated by other people in a constrained environment. The character of the musical material provided by the composer, in conjunction with the particular interactive possibilities afforded to the user, impacts how a structured improvisation with the original musical material is likely to turn out.

These kind of templates also exist in current improvisational practice. Recall composer John Zorn's Cobra mentioned earlier. In this case, the templates are cards and gestures that the prompter may use to choreograph the improvisation. Cobra reflects Zorn's interest in music that has radical and sudden shifts between different styles. One of the cards used in Cobra instructs everyone who is not playing to suddenly start, and everyone who is currently playing to suddenly stop. By making this card available to the prompter, Zorn virtually ensures that any given session of Cobra will have sudden musical changes.

5.5.5   Tonality, Rhythm, and Structure in Mimi4x

By restricting the tonal palette of the materials for Mimi4x's oracles, we can be assured that they will coordinate harmonically. This principle can be observed in Terry Riley's "In C," discussed earlier. In Riley's piece, players are given a series of melodic figures. They are played in order, but repeated an indefinite number of times, such that players eventually go 'out of phase' with one another. The tonal content of the figures are restricted such that they will always sound harmonious one with another. However, the expansive variety generated by the possible overlapping and interaction of materials is enough to sustain interest and attention. While the score of "In C" can be notated on a single page, performances can last for an hour or more. Similarly, several oracles given a set of tonally limited materials can generate an impressive variety of ideas and textures.

Regarding rhythm, repeating a single regular pulse is difficult with multiple oracles, since the four Mimi instances do not communicate with each other, and coordination is essential for a consistent pulse. Thus, a less obvious way of organizing rhythmic structure must be found.

One way is to create 'atemporal' music that denies all sense of pulse altogether. For example, ambient and drone-based music often accomplishes this purpose. Another possibility is to create several different pulses, so that the listener is pulled back and forth between them and must continually re-evaluate the relationship between them. It is ideal if the pulses are related to one another (e.g. 16 note quadruplet to 16 note quintuplet, a 4:5 ratio), so that they sound more compatible. This kind of polyrhythm was first proposed by Henry Cowell (1897–1965) (Cowell 1996) and has recently been referred to (by Kyle Gann and others) as 'totalism' (Gann 2006), due to the complexity generated by this rhythmic activity.

Mimi's factor oracles are already able to build large musical structures through the repetition and recombination of small cells of material. However, a teleological component is missing, i.e. a goal-oriented musical narrative progressing from point A to point B with both a beginning and an end. Since the oracles choose material based on similarity alone, they are agnostic regarding the order of materials and must be started and stopped manually. While this might be undesirable in a concert setting, it is ideal for an installation. The music can continue endlessly without human intervention, while still providing an avenue for human interaction. Users can enter into the work by choosing which oracles will perform and when, and by affecting the recombination rate for each oracle, and the volume of playback. This method of interaction is extremely simple and approachable to all users regardless of their musical background. Yet it also gives the user significant and satisfying influence over the musical structure, since the content of each oracle is distinct and identifiably unique.

5.5.6   The User's Experience with Mimi4x

The user's experience is at the cusp between improvisation and composition. A composer provides the underlying musical ideas; the user is the one who decides which ideas should come on and when, thus taking on some of the traditional roles of the composer. In musical terms, the composer and the oracle determine the foreground (low level) details, while the user makes the background (high level, large structural) to middle-ground decisions, but exercises no control over the note-to-note details. The user also sets the recombination rate of each oracle (a composition decision) and the volume of each music stream (a composition/performance decision) during a Mimi4x session. Thus, when interacting with Mimi4x, the user is in effect improvising the composing of a piece; and, when an audience is involved, this improvised composing becomes a performance as well.

The partnership between user and composer created by Mimi4x is reminiscent of the collaborative relationship between pianist David Tudor (1926–1996) and composer John Cage (1912–1992). Tudor's role as a pianist in performing Cage's music was often more extensive and creatively involved than one might initially suspect. Cage's notation and instructions to the performer were often intentionally ambiguous, leaving room for interpretive musical decisions beyond what is typically expected from a performer. Tudor's realizations of Cage's scores often included engagement in high-level structural decision-making usually reserved for the composer. When Tudor turned to electronic music and took on the role of composer himself, his process was influenced by Cage's use of open forms. In this scenario Tudor viewed his e-equipment as his performing collaborators (Austin et al. 1991). This is analogous to the process of discovery that the user encounters with Mimi4x: by experimenting with the oracles, the user becomes acquainted with their individual personalities.

From a pedagogical perspective, by experimenting with the combination of musical ideas, the user has the opportunity to observe these musical ideas from different perspectives and in different contexts (in different mixes and combinations) and to hear how the ideas interact. The users, general members of the public, thus become directly involved in music making and in reflection on their musical (composition and performance) decisions.

5.6   Conclusion

We have described Mimi, a system for human-computer musical improvisation, and Mimi4x, a system for user-guided structural improvisation. Performing with Mimi is a true creative collaboration in the sense that neither the performer nor Mimi is solely responsible for the content and structure of the performance. Similarly, Mimi4x blurs the boundaries between composition and improvisation, with creative input shared amongst the composer, the four Mimi instances and the user. This raises questions about the ownership of creativity when interacting with digital media.

Returning to George Lewis' conception of creativity in Voyager, he presents composition and improvisation as traditionally oppositional concepts, and it is true that the usual kind of distinctions that people make between the two tend to break down when discussing any kind of computer model of composition or improvisation. What we think of as "spontaneity" in the context of improvisation is difficult to ascribe to a machine, since the decisions that a machine makes tend to be either rule-based or stochastic. Rule-based systems are inflexible with their axioms, while decisions made stochastically must have an element of arbitrariness to them. Whereas, when we think of improvisational spontaneity in people, it includes the flexibility to discard an axiom if it proves to be harmful or inconvenient, and the intentionality to make that decision feel meaningful and coherent.

However, rule-based and stochastic systems are clearly capable of doing things that seem creative, spontaneous, or inspired, as anyone who has received an unexpectedly poetic spam email can attest to. It's worth examining why this is the case. Is it simply a perceptual hack that tricks us into perceiving creativity where there is none? Is human creativity perhaps far more trivial than we imagine? Or do these moments of seeming artificial lucidity reveal something specific, something deeper about human creativity?

Acknowledgement: This material is based in part on work supported by the National Science Foundation under Grant No. 0347988. Any opinions, findings, and conclusions or recommendations expressed in this material are those of the authors and do not necessarily reflect the views of the National Science Foundation.

References

Assayag G, Dubnov S (2004) Using factor oracles for machine improvisation. Soft Comput 8(9):604–610

Assayag G, Rueda C, Laurson M, Agon C, Delerue O (1999) Computer assisted composition at IRCAM: patchwork & openmusic. Comput Music J 23(3):59–72

Assayag G, Bloch G, Chemillier M, Cont A, Dubnov S (2006). Omax brothers: a dynamic topology of agents for improvization learning. In: Proceedings of ACM workshop on music and audio computing, Santa Barbara

Austin L, Boone C, Serra X (1991) Transmission two: the great excursion (TT:TGE): the aesthetic, art and science of a composition for radio. Leonardo Music J 1(1):81–88

Bamberger J (2000) Developing musical intuitions: a project-based introduction to making and understanding music. Oxford University Press, New York

Boivin J (2007) Musical analysis according to Messiaen: a critical view of a most original approach. In: Dingle C, Simeone N (eds) Olivier Messiaen: music, art and literature. Ashgate, Aldershot, pp 137–157

Boulez P (1991) Stravinsky remains. In Stocktakings from an apprenticeship. Trans stephen walsh. Clarendon Press, Oxford, pp 55–110

Bresson, J., C. Agon, and G. Assayag (2005). OpenMusic 5: A Cross-Platform Release of the Computer-Assisted Composition Environment. In 10th Brazilian Symposium on Computer Music, Belo Horizonte, Brazil.

Cowell H (1996) New musical resources, annotated, with an accompanying essay, by David Nicholls. Cambridge University Press, Cambridge

Dubnov S, Assayag G (2005) Improvisation planning and jam session design using concepts of sequence variation and flow experience. Proceedings of international conference on sound and music computing, Salerno

Farbood M (2001) Hyperscore: a new approach to interactive computer-generated music. Master's thesis, Massachusetts Institute of Technology Media Laboratory

François ARJ (2009) Time and perception in music and computation. In: Assayag G, Gerzso A (eds) New computational paradigms for computer music. Editions Delatour France/IRCAM, pp 125–146

François AR, Chew E, Thurmond D (2007) Visual feedback in performer-machine interaction for musical improvisation. Proceedings of international conference on new interfaces for musical expression, New York, pp 277–280

François AR, Chew E, Thurmond D (2011) Performer centered visual feedback for human-machine improvisation. ACM Comput Entertain 9(3):Article 13

Galeyev BM, Vanechkina IL (2001) Was Scriabin a synesthete? Leonardo Music J 34(4):357–362

Gann K (2006) Music Downtown: writings from the village voice. University of California Press, Berkeley

Lévy B (2009) Visualising OMax. Masters Thesis, Ircam/Université Pierre et Marie Curie

Lewis G (2000) Too many notes: computers, complexity and culture in voyager. Leonardo Music J 10:33–39

Pachet F (2003) The continuator: musical interaction with style. J New Music Res 32(3):333–341

Puckette MS (2004) A divide between 'compositional' and 'performative' aspects of Pd. In: Proceedings of the first international pd convention, Graz

Solomon D (2009) In Bb 2.0- a collaborative music/spoken word project. www.inbflat.net

Thom B (2000a) Bob: an interactive improvisational companion. In: Proceedings of international conference on autonomous agents (Agents-2000), Barcelona

Thom B (2000b) Unsupervised learning and interactive jazz/blues improvisation. In: Proceedings of seventeenth national conference on artificial intelligence, Austin

Thom B (2003) Interactive improvisational music companionship: a user-modeling approach. User Model User-Adap Interact J 13(1–2):133–177 (Special Issue on User Modeling and Intelligent Agents)

Walker WF (1997) A computer participant in musical improvisation. Proceedings of human factors in computing systems (CHI), Atlanta

Walker W, Belet B (1996) Applying ImprovisationBuilder to interactive composition with midi piano. Proceedings of the international computer music conference, Hong Kong

Walker W, Hebel K, Martirano S, Scaletti C (1992) Improvisationbuilder: improvisation as conversation. Proceedings of the international computer music conference, San Jose

Weinberg G, Driscoll S (2006) Robot-human interaction with an anthropomorphic percussionist. Proceedings of human factors in computing systems (CHI), Montreal

Zorn J (2004) The game pieces. In: Cox C, Warner D (eds) Audio culture: readings in modern music. Continuum, New York, pp 196–200

Chapter 6
Delegating Creativity: Use of Musical Algorithms in Machine Listening and Composition

Shlomo Dubnov and Greg Surges

6.1 Introduction

In the recent years the ability of computers to characterize music by learning rules directly from musical data has led to important changes in the patterns of music marketing and consumption, and more recently also adding semantic analysis to the palette of tools available to digital music creators. Tools for automatic beat, tempo, and tonality estimation provide matching and alignment of recordings during the mixing process. Powerful signal processing algorithms can change the duration and pitch of recorded sounds as if they were synthesized notes. Generative mechanisms allow randomization of clip triggers to add more variation and naturalness to what otherwise would be a repetitive, fixed loop. Moreover, ideas from contemporary academic music composition, such as Markov chains, granular synthesis and other probabilistic and algorithmic models slowly find their way in, crossing over from experimental academic practices to mainstream popular and commercial applications. Procedurally generated computer game scores such as the one composed by Brian Eno for Maxis' 2008 "Spore" and albums such as Björk's 2011 "Biophilia"—in which a traditional album was paired with a family of generative musical iPad "apps"—are some recent examples of this hybridization (Electronic Arts 2013).

These developments allow composers to delegate larger and larger aspects of music creation to the machines. Accommodating this trend requires developing novel approaches to music composition that allow specification of desired musical outcomes on a higher meta-creative level. With the introduction of such sophisticated software models into the process of music production, we face new challenges to our traditional understanding of music. In music research, new evidence establishes

S. Dubnov (✉) · G. Surges
Music Department, University of California in San Diego,
San Diego, USA
e-mail: sdubnov@ucsd.edu

G. Surges
e-mail: gsurges@ucsd.edu

N. Lee (ed.), *Digital Da Vinci,* DOI 10.1007/978-1-4939-0536-2_6,

connections between computational models of music, cognitive musicology and neuroscience. In order to develop novel paradigms for music that incorporate autonomous or semi-autonomous musical machine behavior as part of the musical creation, we need to better understand the nature of aesthetic music perception on one hand, and practices of composition and improvisation on another.

The basic premise described in this paper is that musical activity is designed to generate pleasure on various perceptual and cognitive levels. By mathematical modeling of behavioral aspects of musical planning, action and decision making, we are able to capture aspects of musical design that were previously intractable in formal ways, and eventually determine the aesthetic aspects of the music itself. Our proposed research is informed by findings of behavioral theories and neurophysiology of reward, reinterpreting past findings about learning, goal-directed approach behavior, and decision making under uncertainty. The mechanisms studied in the reward systems of the brain include dopamine neurons, orbitofrontal cortex, and striatum, mechanisms that can be related to basic theoretical terms of reward and uncertainty, such as contiguity, contingency, prediction error, magnitude, probability, expected value, and variance (Schultz 2006; Salimpoor et al. 2011). As will be discussed below, these features are directly relevant to the study of information in musical signals using a method of analysis called Music Information Dynamics and consequently are also relevant to aesthetic perception and creative design.

In this chapter we will survey methods of statistical music modeling and machine learning of musical style, and relate them to research on computer audition that tries to link human perception, such as emotions, anticipation and aesthetics to quantifiable aspects of musical structure. We also describe methods for automatic generation of new music from learned models of existing pieces.

6.2 History of Mathematical Theory of Music and Compositional Algorithms

The ancient Greek philosopher Pythagoras (570–495 B.C.) is generally credited with having discovered the mathematical relations of musical intervals which result in a sensation of consonance. In Pythagorean tuning, the frequency ratios of the notes are all derived from the number ratio 3:2, which is called an interval of a fifth in music theory. Moving along the circle of fifth results in pentatonic and diatonic scales (taking 5 and 7 steps, accordingly). Although this tuning was adjusted later on in history to accommodate for more complex chromatic musical styles, the idea that music sensations can be related to mathematical proportions and rules had been rooted deeply in music theory.

In terms of creative practices, probably the best known early example of using formal rules for composing music can be attributed to Guido d'Arezzo who, around 1026, set text to music by assigning pitches to different vowels. According to this method a melody was created according to the vowels present in the text. In the fourteenth and fifteenth centuries isorhythmic techniques were used to create repeated rhythmic cycles (talea), which were often juxtaposed with melodic cycles of different length to create long varying musical structures. Dufay's motet Nuper

Rosarum Flores (1436) used repetition of talea patterns at different speeds in four sections according to length ratios 6:4:2:3, which some claim were taken from the proportions of the nave, the crossing, the apse, and the height of the arch of Florence Cathedral or from the dimensions of the Temple of Solomon given in Kings 6:120.

Formal manipulations such as retrograde (backward motion) or inversion (inverting the direction of intervals in a melody) are found in the music of J. S. Bach and became the basis for the twentieth century twelve-tone (dodecaphonic) serial techniques of Arnold Schoenberg (1874–1951). Exploiting the recombinant structure of harmonic progression, Mozart devised his Musikalisches Wurfelspiel ("Musical Dice Game") (1767) that uses the outcomes of a dice throw and a table of pre-composed musical fragments to create a Minuet. There are 176 possible Minuet measures and 96 possible Trio measures to choose from. Two six-sided dice are used to determine each of the 16 Minuet measures and one six-sided die is used to determine each of the 16 Trio measures.

The search for formal compositional techniques took on a life of its own in post-World War II academic music. In a trend which is sometimes attributed to the desire to break away from the stylistic confines and associations of late Romanticism, mathematical rules such as serialism were sought in order to construct musical materials that sounded both new and alien to traditional tonal and rhythmical musical languages. One such feeling was expressed by the Czech novelist Milan Kundera in his writing about the music of the Greek composer and microsound theorist Iannis Xenakis (1922–2001)—"Music has played an integral and decisive part in the ongoing process of sentimentalization. But it can happen at a certain moment (in the life of a person or of a civilization) that sentimentality (previously considered as a humanizing force, softening the coldness of reason) becomes unmasked as the supra-structure of a brutality .... Xenakis opposes the whole of the European history of music. His point of departure is elsewhere; not in an artificial sound isolated from nature in order to express a subjectivity, but in an earthly objective sound, in a mass of sound which does not rise from the hu-man heart, but which approaches us from the outside, like raindrops or the voice of wind" (Kundera 1981).

The use of automation or process is the common thread in the above examples. Composers have increasingly allowed portions of their musical decision-making to be controlled according to processes and algorithms. In the age of computer technology, this trend has grown exponentially. Before going into more technical aspects of musical modeling with computers, it is helpful to compare these experimental academic music approaches to those in the non-academic musical world. To gain this perspective requires briefly surveying the different stylistic approaches for using computers in popular music making.

6.3 Generative Music in Popular Music Practices

One genre that has become popular among the general listening public is ambient music. Ambient music is music which emphasizes mood and texture over directionality and contrast in order to become part of the environment in which it is heard.

This music lacks the dramatic or sentimental design of classical and popular music, but unlike the experimental opposition to tradition of Western music by Xenakis, this genre centers around familiar and often pleasant sounds drawn either from tonal harmonic language or from recordings of environmental sounds that blend into each other with no particular larger scale form or intention.

Perhaps the first explicit notion of ambient music comes from Erik Satie's concept of *musique d'amebulement*, translated as furniture music. Satie's music is often built from large blocks of static harmony, and creates a sense of timelessness and undirected sound, but this was music which was meant not to be listened to, only to be heard as background to some other event (Gallez 1979).

There is arguably no composer more associated with ambient and environmental sound than John Cage, the 20th-century experimentalist who redefined the materials and procedures of musical composition. Throughout his career, Cage explored the outer limits of sonic experience, but his most notorious work is the famous "silent piece", 4'33", composed in 1952. Instead of defining musical ma-terials, Cage defined a musical structure, and allowed the ambient sounds of the concert environment to fill the structure (Pritchett 1996).

The late twentieth century saw an explosion of musical styles which challenged traditional notions of musical experience and experimented with the ways in which sound can communicate space, place, and location. "Soundscape" composition and the use of field recordings help to create a sense of a real or imagined place, while genres of electronic dance music like chill-out are often associated with specific club environments. Some artists have also bridged the gap between con-temporary visual and installation art and music, and "sound art" is now common-place. These artists often deal with psychoacoustic phenomena and relationships between visual and auditory media, often work with multiple forms of media, and take influence from minimalism, electronic music, conceptual art, and other trends in 20th-century art.

Most importantly for our discussion is the way in which composers of ambient and related genres often delegate high-level aspects of their work to algorithmic process or generative procedures—or in the case of field-recording work, to features of the sound materials themselves. Techniques which originated in the most avant-garde and experimental traditions, such as looping, automated mixing, and complex signal processing, are increasingly used in popular music and are now commonly found in commercial, off-the-shelf music production software.

Brian Eno's Ambient 1: Music for Airports (1978) provides the template for many later works: repetitive, with an absence of abrasive or abrupt attacks and us-ing long decays, the harmony and melody seem to continue indefinitely. During the compositional process, Eno constructed tape loops of differing lengths out of various instrumental fragments (Weiner 2013). As these loops played, the sounds and melodies moved in and out of phase, producing shifting textures and patterns. The studio configuration itself became a generative instrument.

In 1996 Brian Eno released the title "Generative Music 1" with SSEYO Koan Software. The software, published in 1994, won the 2001 Interactive Entertainment BAFTA for Technical Innovation. Koan was discontinued in 2007, and the following year a new program was released. Noatikl 'Inmo'—or in the moment ('inmo')—Generative Music Lab was released by Intermorphic ltd. as a spiritual successor to Koan. The software had new generative music and lyric engines, and the creators of the software claimed that

> ...learning to sit-back and delegate responsibility for small details to a generative engine is extremely liberating.... Yes, you can fo- cus on details in Noatikl, but you can also take more of a gardener's view to creating music: casting musical idea seeds on the ground, and selecting those ideas which blossom and discarding those that don't appeal.

Electronic dance music producers, like the English duo Autechre, often embrace the most avant-garde trends and techniques, while still producing music which remains accessible and maintains a level of popular success. Autechre have often employed generative procedures in their music-making, using self- programmed software and complex hardware routings to produce endless variations on and transformations between thematic materials. In a recent interview, Rob Brown spoke of attempting to algorithmically model their own style:

> Algorithms are a great way of compressing your style....if you can't go back to that spark or moment where you've created something new and reverse-engineer it, it can be lost to that moment... It has always been important to us to be able to reduce something that happened manually into something that is contained in an algorithm. (Pequeno 2013)

Commercial digital audio workstations (DAWs)—computer software for recording, mixing, and editing music and sound—often support various types of generative procedures. For example, the "Follow" function in Ableton Live, a popular DAW, allows composers to create Markov chain-like transitions between clips. One of the typical uses of this is to add naturalness to a clip. Instead of having a single, fixed loop, one records multiple variations or adds effects, such as cuts, amplitude envelopes or pitch changes, to a single loop and uses "Follow" to sequence the clips randomly. This allows variations both in terms of the musical materials and expressivity that breaks away from mechanical regularity and synthetic feel often associated with loop-based musical production.

Some other new trends in composing with DAWs can be mentioned:

- Extensive use of effect automation and control curves. The ability to dynamically control effects on individual tracks/ stems has been elevated to the level of compositional practice. Certain musical gestures and evolutions of sound materials are now attributed to the processing phase in the DAW signal chain, instead of traditional composition done by changing the musical materials (notes, recordings) directly.
- Side-chaining and dummy tracks: The idea of side-chaining emerged in French techno as a way to create rhythmic pulsations from otherwise constant synthetic harmonies or "pads". Side-chaining is performed by using a signal compressor to map the amplitude envelope of one sound—often rhythmic—onto another, static sound. This way the rhythmic properties of the first sound are "transferred" to the other track. Today it is possible to use side-chaining on practically any effect (though usually through non-standard plugins). By combing side-chaining with the automation and "Follow" techniques mentioned above, algorithmic or semi-aleatoric methods can now be used to affect any kind of musical material.

- Midi effects: Detailed note or chord manipulation can be performed using MIDI. MIDI plugin effects that generate random notes or melodies can be constrained to adhere to particular scales or chord progressions. These are powerful tools for creating variations and random materials that sound correct within certain tonal frameworks. In other words, computers can be programmed to avoid "false notes" when creating new music according to generative or stochastic procedures.
- Visual maps of musical rules: Additional help with hitting the right notes comes from novel representations of musical scales and chords. New input devices have been designed with intuitive 2D layouts, such as Ableton's "Push" hardware controller and Migamo's "Polyplayground" iPad app. These devices move away from the traditional piano keyboard, and instead exploit various geometric representations of scales and pitch spaces. Ableton Live now integrates with Max/MSP, through Max For Live, and allows user-created additions and modifications to the Live production environment.

Computer and video games more and more frequently make use of generative or procedural music and audio. Advantages of procedural audio include interactivity, variety, and adaptation (Farnell 2007). Karen Collins provides a good overview of procedural music techniques (Collins 2009). Collins describes the challenges faced by game composers, who must create music which both supports and reacts to an unpredictable series of events, often within constraints imposed by video game hardware and/or storage. In fact, Collins points out that a contributing factor to the use of generative strategies in video games was the need to fit a large amount of music into a small amount of storage. Generative techniques allow for the creation of large amounts of music with smaller amounts of data. Some strategies for this kind of composition can include:

- Simple transformations: changes in tempo, restructuring of largely precomposed materials, conditional repeats of material based on in-game activities.
- Probabilistic models for organizing large numbers of small musical fragments.
- Changes in instrumentation of music based on the presence of specific in-game characters.
- Generation of new materials within melodic/rhythmic constraints.

6.4 Computer Modeling of Music

On August 09, 1956 the "Illiac Suite" by Lejaren Hiller and Leonard Isaacson, named after the computer used to program the composition, saw its world premiere at the University of Illinois (Bewley 2004). The work, though performed by a traditional string quartet, was composed using a computer program, and employed a handful of algorithmic composition techniques. This work, together with the work of Xenakis described in *Formalized Music*, marks the beginning of modern mathematical and computational approaches to representing, modeling and generating music. Among the most prominent algorithmic composition techniques are stochastic or random modeling approaches, formal grammars, and dynamical systems that consider fractals and chaos theory, as well as genetic algorithms and cellular automata for generation and evolution of musical materials. More recently, advanced machine learning techniques such as neural networks, hidden Markov models, dynamic texture models and deep learning are being actively researched for music retrieval, and to a somewhat lesser extent, also for creative or generative applications.

Markov models are probably one of the best-established paradigms of algorithmic composition. In this approach, a piece of music is represented by a sequence of events or states, which are symbolic music elements such as notes or note combinations, or sequences of signal features extracted from a recording. Markov chains are statistical models of random sequences that assume that the probability for generating the next symbol depends only on a limited past. A Markov model establishes a sequence of states and transition probabilities, which are extracted by counting the statistics of events and their continuations from a musical corpus. The length of the context (the number of previous elements in the musical sequence used for the calculation of the next continuation) is determined by the order of the Markov model. Since higher order models produce sequences that are more similar to the corpus, but are harder to estimate and may also have disadvantageous effects in terms of the model's ability to deal with variations, several researchers have turned their attention to predictive models that use variable memory length models, dictionary models and style specific lexicons for music recognition and generation. See (Conklin 2003) for a survey of music generation from statistical models and (Dubnov 2006) for comparison of several predictive models for learning musical style and stylistic imitation.

Generative grammars are another well-established and powerful formalism for the generation of musical structure. Generative grammars function through the specification of rewriting rules of different expressiveness. The study of grammars in music in many ways paralleled that of natural language. Chomsky erroneously believed that grammatical sentence production cannot be achieved by finite state methods, because they cannot capture dependencies between non-adjacent words in sentences or model the recursive embedding of phrase structure found in natural language (Chomsky 1957). Since such complex grammars could not be easily "learned" by a machine, this belief limited the types of music handled by grammar models to those that could be coded by experts in terms of rules and logic operations. This approach is sometimes called an expert system or knowledge engineering. Although these methods achieve impressive results, they are labor intensive as they require explicit formulation of musical knowledge in terms that are often less than intuitive.

Some more specific models, such as Transition Networks and Petri Nets, have been suggested for modeling musical structure. These are usually used to address higher levels of description such as repetition of musical objects, causality relations among them, concurrency among voices, parts and sections and so on. David Cope's work describes an interesting compromise between formal grammar and pattern-based approaches. Cope uses a grammatical-generation system combined with what he calls "signatures": melodic micro-gestures typical of individual composers (Cope 2004). By identifying and reusing such signatures, Cope re-produced the style of past composers and preserved a sense of naturalness in the computer-generated music.

In this context it is worth also mentioning recent work on music representation that was summarized under a new standard for music and multimedia encoding called IEEE 1599 (Baggi and Goffredo 2009). This standard has been developed to allow interaction with music contents through the use of six layers that combine prerecorded contents with structural and logical representations to allow interactive access within the compound piece. Petri Nets are used in the Structural Layer for identifying music objects and their relationships, thus allowing representation of different kinds of musicological analyses. The other layers include the General Layer containing catalog information, Logic Layer that provides the symbolic description with timing and synchronization information, and Notational, Performance and Audio Layers that contain pre-recorded clips in multiple formats. It should be noted that very much like other formal language approaches, encoding music into this format is quite demanding in terms of human involvement.

In the following we will describe some of the research on learning musical structure that begins by attempting to build a model by discovering phrases or patterns which are idiomatic to a certain style or performer. This is done automatically, and in an unsupervised manner. This model is then assigned stochastic generation rules for creating new materials that have similar stylistic characteristics to the learning example. One of the main challenges in this approach is formulating the rules for generation of new materials that would obey aesthetic rules or take into account perceptual and cognitive constraints on music making and listening that would render pleasing music materials.

6.5 Machine Improvisation

Another interesting field of research is machine improvisation, wherein a computer program is designed to function as an improvisational partner to an instrumentalist. The software receives musical input from the instrumentalist, in the form of MIDI or audio, and responds appropriately according to some stylistic model or other algorithm. An important early work in this style is George Lewis's *Voyager* (1988), a "virtual improvising orchestra." The *Voyager* software receives MIDI input derived from an instrumental audio signal, analyzes it, and produces output with one of a set of many timbres (Lewis 2000). The input data is analyzed in terms of average pitch, velocity, probability of note activity and spacing between notes. In response, *Voyager* is capable of both producing variations on input material and generating completely new material according to musical parameters such as tempo (speed), probability of playing a note, the spacing between notes, melodic interval width, choice of pitch material based on the last several notes received, octave range, microtonal transposition and volume. Due to the coarse statistical nature of both the analysis and generative mechanisms, the type of music produced by the system is limited to abstract free improvisation style.

OMax and SoMax, which will be discussed in greater detail below, form part of another family of "virtual improvisor" software. These programs, along with our own PyOracle, utilize sequence matching based on an algorithm called Factor Oracle, to create a model of the repetition structure specific to the incoming music material both for audio or MIDI input. By recombining carefully selected segments of the original recording according to this repetition structure, the programs can produce new variations on the input, while maintaining close resemblance to the original musical style.

6.6   Modeling of Musical Style

The current line of research started with the well-known Lempel-Ziv (LZ) coding technique (Ziv and Lempel 1977), which takes a series of symbols from a finite alphabet as input, and builds a tree of observed continuations of combinations of the input symbols. This tree grows dynamically as the input is parsed using what is called the incremental parsing (IP) method. If the input is a sample of a stationary stochastic process, LZ asymptotically achieves an optimal description of the input in the sense that the resulting tree can be used to encode the input at the lowest possible bit rate (the entropy of the process). This implies that the coding tree somehow encodes the law of the process; for instance, if one uses the same tree (as it stands after a sufficiently large input) to encode another string, obeying a different law, the resulting bit rate is higher than the entropy.

This fact can be used to test whether a new sample is likely to have arisen from the original process or not. Moreover, one can use the decoding algorithm on a totally random process of codeword selection; this produces samples of the process, or—in musical terms—it creates variations or improvisations on the same musical information source, at least in the sense that the re-encoding test will fail to distinguish it from a "genuine" sample (Dubnov and Assayag 2002). This information theoretic approach to music gave rise to many investigations: first, the method has been evaluated on specific musical styles, using different choices of alphabet and approaches to polyphony. Second, although LZ is asymptotically optimal, many possible variants exist as well, such as some of the more recent results described in the section on music improvisation. Third, many researchers and teams are now applying these techniques for real-time improvisation and composition by computers. In (Dubnov et al. 2003) we have augmented this method by developing musical style analyzers and generators based on Variable Markov Models (VMM). The class of these algorithms is large and we focused mainly on two variants of predictors—universal prediction based on Incremental Parsing (IP) and prediction based on Probabilistic Suffix Trees (PST). Both IP and PST build tree structures in the learning stage, where finding the best suffix consists of walking the tree from the root to the node bearing that suffix. The main difference between the methods is that while IP operates in the context of lossless compression, cleverly and efficiently sampling the string statistics in a manner that allows a compressed representation and ex-act reconstruction of the original string, PST is a lossy compression model that retains only partial information considered essential for prediction and classification.

Fig. 6.1 Suffix links and
repeated factors

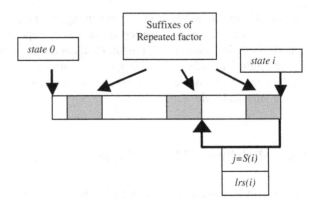

Although many researchers have proposed computational approaches to encoding
musical style, some of which are highly sophisticated and implemented by large
software systems, the information theoretic approach is mathematically elegant and
scientifically testable, which partially explains the interest it has attracted among
computer music researchers.

6.6.1   Improvisation as an Information Generation Process

Computer improvisation within a stylistic framework requires capturing the
idiomatic means of expression of a composer or a genre in attempt to endow
the computer with both creative and "communicative" human-like capabilities.
As such, an improvisation system consists of two main components: a learning
algorithm that builds a model of music sequences in a specific style, and a generation algorithm that operates on the model and produces a new musical stream in
that particular style, so that the next musical unit is selected in a manner that best
corresponds as a continuation for the already generated materials. In real-time applications, the system also must alternate between the learning or listening phase
and the generating/improvising phase. This alternation can be seen as machine
improvisation, where the machine reacts to other musician's playing, after listening to and analyzing his performance. As IP and PST build tree structures in the
learning stage, finding the best suffix involves walking the tree from the root to
the node bearing that suffix. We explored the use of Factor Oracle (FO) proposed
by M. Crochemore for the purpose of sequence generation in a particular style
(Allauzen et al. 1999). The use of FO for generation is shown in Fig. 6.1 below.
In order to use the automation for generation, another set of links $S(i) = j$, called
Suffix Links running backward are used. These links point to node j at the end of
the longest repeating suffix (also called repeating factor) appearing before node i
(i.e. longest suffix of prefix of i that appears at least twice in prefix of i).

FO automation provides the desired tool for efficiently generating new sequences of symbols based on the repetition structure of the reference example. Compared

to IP and PST, FO is structurally closer to the reference suffix tree. Its efficiency is close to IP (linear, incremental). Moreover, it is an automaton, rather than a tree, which makes it easier to handle maximum suffixes in the generation process. This method also differs from Markov chain-based style machines mentioned above in the extent of the signal history or context that it is able to capture.

6.7   Musical Semantics

Issues of musical semantics are usually debated between so called referentialist and absolutist points of view. The referentialist view claims that musical meaning comes either from references to objects outside music, while the absolutist view espoused by the composer Igor Stravinsky, argues that the meaning of music, if any, lies in the music itself and the relations entertained by the musical forms. Compromise between these two views was presented as early as 1950 by Meyer who formed a theory of expectations which brings emotion and meaning to music (Meyer 1956). Meyer used psychologically-based arguments to claim that

> The meaning of music lies in the perception and understanding of the musical relationships set forth in the work of art [thus being primarily intellectual] and that these same relationships, are in some sense capable of exciting feelings and emotions in the listener.

These ideas were later elaborated by Narmour using the notion of expectation, correlating meaning to the act of establishing musical expectation, only to be later deceived or violated. Without committing to any particular theory, these works seem to suggest the general principle that access to music requires association among musical structures and possibly choosing among a multitude of interpretations and representations of meanings. The central challenge in this research is discovering what aspects of musical structure actually have the evidence of eliciting musical experience.

6.7.1   Theories of Auditory Anticipation

More recently, Huron has extended this line of inquiry into a general theory of expectation that involves multiple functionally distinct response systems on different time scales of musical structure (Huron 2006). Being able to discover musical structure on different time scales and make correct short- and long-term predictions or explanations is essential for music understanding and experience. The anticipatory paradigm of music perception also suggests close relations between the prediction and repetition aspects of music and the aesthetic possibilities afforded by expectation including physiological psychology of awe, laughter, and more.

In attempt to provide a formal computational approach to these ideas, an information theoretic construct called information rate (IR) was proposed as a tool to make local maps within an audio recording of the rising and falling rate of musical novelty on different time scales (Dubnov 2006, 2008). This necessitates operating not on the raw audio signal but on dimensionally reduced, psycho-acoustically aware frame-based analyses. This approach has been empirically validated in an experiment with subjects that compared the information rate measure with the measured ebb and flow of musical engagement in the audience during a performance of "Angel of Death", a work composed by Roger Reynolds, which was designed with this experiment in mind (Dubnov et al. 2006). This work is central to our research in the field of machine understanding of music because, for the first time, a class of algorithms is proposed that can deduce musical structure from audio recordings without the aid of a musical score. Recently, the method for measuring information rate was extended towards using the powerful Audio Oracle representation that had been previously used successfully for machine improvisation (Dubnov et al. 2011). This new direction links style modeling to research on anticipation, and opens new possibilities for constructing listening agents that are capable of interactive or autonomous improvisation and aesthetic decision-making.

It should be noted that since the 1950s, many techniques have been proposed to automatically analyze musical scores, but it remains a major challenge to find musical structure in an audio recording. Moreover, the question of formally translating music or sound representation into aesthetic specifications has not been addressed before. The generic nature of our approach that applies both to symbolic and audio data is also important because many musics are either not written in Western music notation or not written down at all; and also because musical performances, even of the classical Western repertoire, contain much information that is not explicitly specified in the score at all and was heretofore not available for machine analysis. Working directly from the audio signal makes such performance related information available to the algorithm.

6.8   Computational Aesthetics

Eighty years ago, the physicist and mathematician Birkhoff formalized the notion of aesthetics in terms of the relations between order and complexity (Birkhoff 1933). This idea was later reformulated in terms of information theory by Bense, who suggested that "complexity" is measured in terms of entropy, and that "order" is the difference between uncompressed and compressed representations of the data (Bense 1969). In the case of music, past musical events can be used for compression. Our method of modeling musical anticipation resembles the computational aesthetics approach, as it provides a graph of order (where order = uncompressed complexity—compressed complexity). This serves as motivation to develop methods for estimating these "temporal aesthetics"—measures for explaining the pleasure or fun the audience derive during a listening process. Moreover, an algorithmic composer should be equipped with similar sensibilities in order to guide him through the choice of materials, either autonomously or according to user input in our meta-composition design approach.

6.8.1 Theory of Reward in Aesthetics

Historically, the leading tradition in psychological research on aesthetics has been Daniel Berlynes psychobiological model, as represented by the new experimental aesthetics movement of the 1970s. That theory mostly emphasized the arousal- modifying collative properties of art by identifying factors such as complexity, novelty and uncertainty. More recently, appraisal theories have provided new perspectives on emotional responses to art, bringing to the forefront of research aspects of interest and enjoyment, and also informing other theories (e.g., prototypicality models, fluency and so on). Emotions are grouped according to their appraisal structures. In this approach, rather than attributing features like complexity, novelty, etc., to the stimulus object itself, the viewer appraises the stimulus as having these features. Interest is a central contributing emotion when considering aesthetic interest, and consists of two components. The first is a novelty check, and the second is a coping-potential check. Increasing coping-potential by giving foreknowledge was shown to increase interest (Silvia 2005).

Leder et al. propose a model of aesthetic experience involving perceptual analyses, implicit memory integration, explicit classification, cognitive mastering, and evaluation (Leder et al. 2004). This model results in aesthetic judgment, which is evaluation of the cognitive mastering stage, and aesthetic emotion, which is a by-product of the earlier stages. It is argued that aesthetic experience is largely depending on aesthetic attitude, which determines 'information processing of aesthetic stimuli.' This attitude is defined by pre-existing context, although it is also possible for aesthetic experiences to influence the attitude itself.

- Perceptual analyses: Previous work has shown that small amounts of con-trast along a single dimension can affect aesthetic judgment; that visual complexity is important relative to arousal potential, with preference for a moderate level; that symmetry is detected early and preferred over non- symmetry; and that grouping and 'order' variables are preferred.
- Implicit memory integration: Familiarity and repetition have been generally shown to be important in aesthetic preferences, although this effect is reduced or eliminated by novelty and complexity in a previously-familiar stimulus; prototypicality is preferred; and there is a peak-shift effect in which perceivers respond more strongly to objects that exaggerate characteristic qualities of familiar objects ('art as a search for essential features').
- Explicit classification: Conscious and deliberate analyses are made in regards to content and style; there is processing of stylistic information and the pleasure of generalization; there is recognition of alienation into a novel style.
- Cognitive mastering: This stage involves semantic associations and episodic memory; knowledge of styles and conscious understanding of distinctive features; and other top-down knowledge such as expertise and integration of external information ('elaborate titles,' etc.).

After evaluation and integration of all of these stages, artwork may generate a perceptually pleasure-based or cognitive-based reception. Since aesthetic experience may involve either or both, this complicates empirical work when it comes to test such aesthetic processing theories. This is closely related to functions of rewards that are based primarily on their effects on behavior and are less directly governed by the structure of the input events as in sensory systems. Therefore, the investigation of mechanisms underlying reward functions requires behavioral theories that can conceptualize the different effects of rewards on behavior. The investigation of such processes is done by drawing upon animal learning and economic utility theories (Schultz 2006). This approach offers a theoretical framework that can help to elucidate the computational correlates for reward functions related to aesthetics features that are derived from computational models and information theoretic analysis of musical data.

6.8.2   Information Theory Models in Arts and Entertainment

Some interesting research has been done on applying information theory measures to aesthetic experience. As suggested above, by considering order as the amount of redundancy in a stimulus, and complexity as the amount of randomness, one can begin to consider the use of a compression algorithm to measure an aesthetic value of a signal. Of course such measure depends on the abilities of the compressor to detect and exploit redundancies in the signal, which makes this data beautiful "in the eyes" of the compression algorithm. For instance, by considering a painting as an information source, the aesthetics of different paintings can be measured according to different compression methods applied to them (Rigau et al. 2008). Order can be considered in terms of similarity between sections of an image or encoding of spatial frequencies in optimal ways, and so on. Information theoretic measures have also been applied to television programs (Krull et al. 1977). Different measures of entropy like set time entropy, set incidence, entropy, verbal time entropy, etc. were applied to television shows. Structural measures like unity, variety, pace, etc. were also tested. The variables were factored/combined into a DYNUFAM index, standing for dynamics and unfamiliarity. It was found that age has an influence on preference for structure and entropy in television viewing, with a peak in the middle 20s. Research into auditory structure also found that increases in complexity correspond to increases in listener arousal, measured both by participant reporting and skin conductance meters (Potter and Choi 2006). It was also found that the 30-second point was significant in that structure/complexity began to have a larger effect on the listeners after that time. A somewhat similar approach is to measure the fractal dimension, D, of an image (Spehara et al. 2003). Fractals in nature exhibit a variety of measurements for D, but it was found that D between 1.1 and 1.5 was considered most appealing. By measuring preference for images from nature, computer graphics, and abstract expressionist paintings, all at varying levels of D, similar preference peaks relative to D were shown. Burns suggests a Bayesian model for understanding the perceptual process over time (Burns 2006). Expectation (E) relates to the likelihood of an event given the probabilities of previous events. Violation is the inverse of this expectation. Explanation (E') gives the amount of violation which can be explained. Pleasure and tension can both be computed from EVE', given personality dependent scaling factors. We will return to this model later on in relation to musical expectation.

6.9   Audio Oracle

Audio Oracle (AO) is an analysis method and a representation of musical structure that extends the machine improvisation methods that used LZ or PST, as described above (Dubnov et al. 2007). The AO algorithm is based on a string matching algorithm called Factor Oracle, extending it to audio signals. AO accepts a continuous (audio) signal stream as an input, transforms it into a sequence of feature vectors and submits these vectors to pattern analysis that tries to detect repeating sub-sequences (or factors) in the audio stream. Mathematically speaking, the AO generalizes the FO to partial or imperfect matching by operating over metric spaces instead of a symbolic domain. One of the main challenges in producing an AO analysis is determining the level of similarity needed for detecting approximate repetition. As we will discuss below, this is done using information theoretic considerations about the structure of the resulting AO. In other words, the "listening mechanism" of the AO tunes itself to the differences in the acoustic signal so as to produce a optimal representation that is the most informative in terms of its prediction prop- erties. Like the FO, AO outputs an automaton that contains pointers to different locations in the audio data that satisfy certain similarity criteria, as found by the algorithm. The resulting automaton is passed later to an oracle compression module that is used for estimation of Information Dynamics, as will be described in the following section.

Algorithms 1 and 2 demonstrate pseudo-codes for Audio Oracle construction. During the online construction, the algorithm accepts audio frame descriptors (user-defined audio features) as vectors σi for each time-frame i and updates the audio oracle in an incremental manner. Algorithm 1 shows the main online audio oracle construction algorithm.

Algorithm 1 On-line construction of Audio Oracle

Require: Audio stream as $S = \sigma_1 \sigma_2 \cdots \sigma_N$
1: Create an oracle P with one single state 0
2: $S_P(0) \leftarrow -1$
3: **for** $i = 0$ to N **do**
4:    $Oracle\,(P = p_1 \cdots p_i) \leftarrow$ Add-Frame $(Oracle\,(P = p_1 \cdots p_{i-1}), \sigma_i)$
5: **end for**
6: **return** Oracle $(P = p_1 \cdots p_N)$

Algorithm 1 calls the function Add-Frame—described in algorithm 2—which updates the audio oracle structure using the latest received frame descriptions. This function works very similar to Factor Oracle except that (1) it accepts *continuous* data flow rather than symbolic data, (2) does not assign symbols to transitions (instead each state has a one-to-one correspondence with frames in audio buffer), and (3) it uses a distance function along with a threshold θ to assess the degree of similarity between frame descriptions. The set of links in Audio Oracle are forward transitions $\delta(i, \sigma)$ and suffix links $Sp(k)$. Similar to the Factor Oracle algorithm, forward transitions correspond to states that can produce *similar patterns* with alternative continuations by continuing forward, and suffix links correspond to states that share the *largest similar sub-clip* in their past when going backward.

Algorithm 2 Add-Frame function: Incremental update of Audio Oracle

---

Require: Oracle $P = p_1 \cdots p_m$ and Audio Frame descriptor vector σ

1: Create a new state $m + 1$
2: Create a new transition from m to $m + 1$, $\delta m, \sigma = m + 1$
3: $k \leftarrow S_P(m)$
4: **while** $k > -1$ **do**
5:     Calculate distances between σ and S
6:     Find indexes of frames in S whose distances from σ are less than θ
7:     **if** There are indexes found **then**
8:         Create a transition from state k to $m + 1$, $\delta(k, \sigma) = m + 1$
9:         $k \leftarrow S_P(k)$
10:     **end if**
11: **end while**
12: **if** $k = -1$ (no suffix exists) **then**
13:     $S \leftarrow 0$
14: **else**
15:     $S \leftarrow$ where leads the *best* transition (min. distance) from k
16: **end if**
17: $S_{p\sigma} \leftarrow S$
18: **return** Oracle $P = p_1 \cdots p_m \sigma$

6.9.1 Music Information Dynamics Using Audio Oracle

Music Information Dynamics is a field of study that considers evolution in information contents of music (Dubnov et al. 2006; Potter et al. 2007; Abdallah and Plumbley 2009) which is assumed to be related to structures captured by cognitive processes such as forming, validation and violation of musical expectations (Meyer 1956). In particular, a measure called Information Rate (IR) was studied in relation to human judgments of emotional force and familiarity (Dubnov et al. 2006). Information Rate (IR) measures the reduction in uncertainty about a signal when past information is taken into account. It is formally defined as mutual information between past $x_{past} = \{x_1, x_2, \ldots, x_{n-1}\}$ and the present x_n of a signal x.

$$IR(x_{past}, x_n) = H(x_n) - H(x_n \mid x_{past}) \tag{6.1}$$

with $H(x) = -\Sigma P(x) log_2 P(x)$ and $H(x \mid y) = -\Sigma P(x, y) log_2 P(x \mid y)$ being the Shannon entropy and conditional entropy respectfully, of variable x distributed according to probability $P(x)$. In (Dubnov et al. 2011) an alternative formulation of IR using AO was developed, which we term Compression Rate (CR)

$$CR_{AO}(x_{past}, x_n) = C(x_n) - C(x_n \mid x_{past}) \qquad (6.2)$$

with $C(\cdot)$ being the coding length obtained by a compression algorithm, and measured in terms of the number of bits required to represent each element x_n using a Compror algorithm (Lefebvre and Lecroq 2001).

In order to use AO for compression, the length of a longest repeated suffix needs to be calculated. This is approximated through a quantity called *LRS* that recursively increments the length of the common suffix between two positions in the signal—the immediate past of the current frame and immediate past at a location pointed to by the suffix link. When a new frame $(i+1)$ is added, the suffix link and *LRS* for this frame are computed.

6.9.2   Kolmogorov Dynamics

In using compression algorithms to trace the dynamics of information, a distinction needs to be drawn between statistical compression that uses many samples from the same random source, and so called compression of "individual sequences" that does not require an underlying statistical model. These issues appear for instance in the classical LZ78 paper (Ziv and Lempel 1978) where the authors define a different concept of compressibility and actually prove that their method is asymptotically optimal, i.e. it performs as good as an entropy coder if we knew how to construct it. Unfortunately there is no such proof for the Compror algorithm that we are aware of, although comparison benchmarks have shown that it is comparable and sometimes better then LZ, empirically speaking. Another similar discussion happens in the image aesthetics paper (Rigau et al. 2008). There the authors bypass the problem of dealing with image statistics by referring to compression algorithms as approximations to Kolmogorov or algorithmic complexity. Although this does not resolve the conceptual issue, it conveniently moves the problem to a different domain where the determining factor of the signal complexity is in the hands (or "in the eyes") of the compression program. In an ideal world, such compression would be optimal for the class of signals it operates on, but in practice the coder, such as JPEG or PNG in the visual case, are predetermined. The compression can be adaptive and learn the statistics "on the fly", such as in the case of LZ, but this optimality does not reveal itself from start and is manifested in poor compression of short sequences. This distinction between complexity being intrinsic to the source, i.e. the music or the image itself, or determined by the receiver (a human listener or a computer audition program), is at the core of many debates on aesthetics. Though we do not resolve it here, at least we spell out the conditions and the experimental design considerations pertinent to this matter.

Going back to some more technical details, essentially, a string's Kolmogorov complexity is the length of its ultimate compressed version that is machine independent up to an additive constant. More precisely, the Kolmogorov complexity of a string x is the length of the shortest program to compute x on an appropriate universal computer. So analogously to conditional entropy, the conditional complexity of x relative to y can be defined as the length of the shortest program to compute x given y as a side information available to the algorithm. Accordingly, we generalize here the original IR to include the concept of algorithmic mutual information between x and y as the difference between the compressed length of x alone and the its length when y is available to the compression algorithm. Moreover, since we are interested only in compression of the present given the past, we call it "Compression Rate" analogously to IR. Last remark due is that our compression scheme does not accomplish the ideal terms that Kolmogorov complexity requires, so we treat this as an approximation.

6.9.3 A Simple Example

As a simple example of using compression for estimation of CR in the symbolic case, the word *aabbabbabbab* will be encoded as follows: if a suffix link to a previous location in the sequence is found and the length of the repeating suffix is smaller than the number of steps passed since the last encoding event, then the whole preceding segment is encoded as a pair (length of string to be recopied, position to start recopying from). Accordingly, in our string example, the first letter will be encoded using the letter a using 1 bit over an alphabet $\{a,b\}$. The next occurrence of a can be encoded as a pair (1, 1), but since encoding it will take more bits then encoding a individually, we will use the shorter encoding and proceed to encode the next b individually and then deciding between representing the following b as a pair (3, 1) or as a single letter, choosing the latter option. The compression advantage appears for the remaining portion of the string. According to the encoding method this segment can be encoded as a pair (8, 2), which will take $log_2(8) + log_2(12) = 6.58$ bits, practically offering 1 bit saving compared to encoding of the 8 characters individually.

In the AO case, the symbols are replaced by feature vectors, and repetitions are found by approximate matching between those features up to a certain similarity threshold. From the resulting AO structure we can find the points where the compression occurs according to the method described above. For every frame that occurs between the encoding events we compute the compression gain in terms of the number of bits required to represent the pair (*length, position*), divided by the length of the encoded segment. Since each AO for a specific threshold value can exhibit a different structure, the CR measurement can be used to select the optimal AO by selecting the one having the highest total CR.

Turning to an audio analysis, below are some examples of a code created by the Compror compression algorithm applied to AO.

[(0, 1), (3, 1), (0, 5), (0, 6), (0, 7), (1, 6), ... (1, 134), (0, 173), (29, 173), (3, 201), (2, 100), (0, 208),... (0, 238), (0, 239), (3, 239), (0, 243), (0, 244), (0, 245)]

The music analyzed here is a recording of the fourth movement of Sergei Prokofiev's "Visions Fugitives" (Prokofiev 1955). We generated a good AO representation using a distance threshold of 0.21, analysis frames of 16384 samples (approximately 372 ms) and using Mel-Frequency Cepstral Coefficients (MFCCs) as our audio feature. Shown above are three sections from the codebook, one from the beginning of the recording, one towards the middle and one at the end. A graph showing the CR analysis of this piece is shown below in Fig. 6.10. The reading of the code reveals the basic properties of the AO representation and the grouping done by the Compror method. Each entry represents a pair of (number of frames in a sound clip to be recopied, recopying position). When a new frame arrives that cannot be recopied, we mark it as zero in the first number of the pair. In such case the second number represents the location of a new sound instance in the original file. As we do not actually use this method for compression, the pairs should be regarded as ways to "understand" the musical structure. More discussion of this follows below.

Returning to our sequence, we see that the first sound frame is new, and is then followed by three repetitions. The fifth, sixth, and seventh frames are also new. The eighth is a recopy of the sixth. Following the same syntax, we see longer recopied blocks in the middle of the piece, some of which are immediate repetitions of a new sound, others of which are repetitions from earlier in the piece. It is also worth noticing that the tendency to recopy longer blocks increases in time, but towards the end of the recording three new instances occur, which many time can be due to public applause or even a reverberation tail that was not heard before.

6.10   Music Experience of Expectation and Explanation

Burns proposes an interesting interpretation of aesthetic experience that elaborates the anticipation approach by adding a posterior step of explanation. His theory, called EVE', is based on three basic aesthetic components: expectations (E), violations (V), and explanations (E') (Burns 2006). Burns compares the relations between the IR approach, in its information theoretic formulation, and the Bayesian view of EVE'. It is beyond the scope of this paper to survey the relations between the two models, but for our purposes it would be sufficient to say that the ability to explain an object or make sense of a musical stream seem to be natural elements in our enjoyment of music. In Burn's approach, explanation is the discrepancy between models that occurs after we have seen the object or listened to music. It is closely related to Baldi and Itti's model of Surprisal that considers an event to be surprising if the explanations (i.e. the underlying models, sometime also called latent parameters) change significantly before and after observing new data (Itti 2005). More-over, Huron includes an a-posteriori aspect in his Imaginative-Tension-Predictive-Reactive-Appraisal (ITPRA) theory of anticipation. According to Huron, the first two steps in forming expectations occur prior to the onset of a stimulus, while the three remaining steps are post-outcome and are related to fast unconscious and slower reactive responses that require learning and appraisal of the reactive reflex. Finally, maybe the best justification for explanation E' can be found in Aristotle's reversal-recognition idea, where he writes: A whole is that which has a beginning, a middle and an end (p. 31). In between these three parts are two paths, which he calls complication (beginning-middle) and unraveling (middle-end). Aristotle explains that an effect is best produced when the events come on us by surprise; and the effect is heightened when, at the same time, they follow as cause and effect (p. 39). This is why deus ex machina ("god from the machine") are considered poor plot devices as they provide interventions to a seemingly unsolvable problem in ways that deny explanation or cause-and-effect[1]. However, one should be careful in employing or giving too much significance to the names we use to denote the different phases of aesthetic perception, as our language seems to be poorly fitted to distinguish between the fine details of the different modeling approaches. For the purpose of the current discussion, we will consider expectations as a zero Markov order entropy. In other words, the effort required to encode a sequence of musical events when no memory is employed to find repeating structures (beyond instantaneous recognition of a sound object as belonging or new to some set of sounds heard before). The explanation aspect in our model will be considered as an ability to recognize a segment of music and link it to a prior occurrence of a similar segment. We can draw a fairly straightforward mapping between the encoding methods presented in the previous section and EVE'. Since V is considered in Burn's theory as negative E, and is omitted from actual calculations, we will be using E and E' as follows:

- Expectation (E) will be measured in terms of the growth in the size of the alphabet (i.e. the number of new sounds) that occur over time during a listening process.
- Explanation (El) is measured in terms of the saving in coding length when links to earlier occurrences of sound were found, as represented by the AO compression pairs.

The total experience of pleasure, fun or flow F (these terms are used inter-changeably in (Burns 2006) is a weighted sum of E and E', written as $F = G * E + G' * E'$, where weights G and G' are set using various heuristic considerations. In our case we will measure E as entropy H_0 and E' as H_1 and the flow is equated to IR (or CR for that matter) as $IR = H_0 - H_1$. The pseudo-code for estimating H_0, H_1 and CR, (or E, E' and F, respectfully) is:

This algorithm counts the size of the alphabet up to moment i both with and without using information about similar blocks in the past. Whenever a new and

---

Algorithm 3 Incremental CR from AO compression code

Require: A sequence of codeword pairs ($L = length, location$) and total signal length N

1: Create counters H_0 and H_1
2: **for** $i = 1$ to N **do**
3:    $H_0 \leftarrow Log_2$ (number of new states ($L == 0$) up to i)
4:    $H_1 \leftarrow \dfrac{Log_2 \text{ (number of all codewords up to } i)}{\text{length L of a block to which state } i \text{ belongs}}$
5: **end for**
6: **return** $CR = H_0 - H_1$

---

[1] The name comes from a crane (mekhane) that was used to lower actors playing gods onto the stage to resolve the plot.

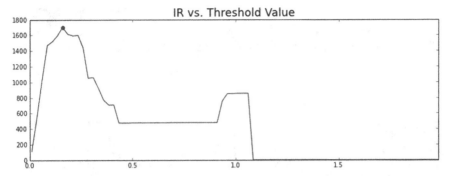

Fig. 6.2 Evolution of total CR as a function of distance threshold

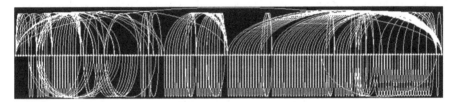

Fig. 6.3 Optimal Audio Oracle structure built on the fourth movement of Sergei Prokofiev's "Visions Fugitives"

distinct sound frame is encountered ($L = 0$), the alphabet increases by one more element. The number of bits needed to encode an alphabet of size Q is $Log_2(Q)$. This is also the upper limit to the entropy of encoding a sequence of drawn from a set of Q possible outcomes. Accordingly, as time passes, more and more new states (frames) are encountered, and the size of the alphabet increases. In our interpretation, this is the upper limit to the entropy of the signal up to moment i, which represents the "Expectation" part of the Fun equation. The "Explanation" part is considered in terms of the ability to trace links to the past (explaining the present via the past) and is measured as the coding length using the AO compression scheme. In this case, a new codeword appears with every encoded block, but the advantage in terms of coding length is that a long block of sound frames is captured by a single Compror pair. Accordingly, the number of bits required to encode any individual frame is the number of bits needed to encode the codeword divided by the length of the block (Fig. 6.2).

For a more intuitive understanding of CR and how it relates to the AO structure, consider Figs. 6.3–6.5. These figures demonstrate the difference between an oracle constructed using a well-chosen distance threshold and one constructed where the threshold is too low or too high. The first, shown in Fig. 6.3, is considered the optimal oracle. Through an adaptive process, we compare the total CR of each of a range of possible distance thresholds, and select the particular threshold which maximizes total CR. Figure 6.2 shows the total CR as a function of the distance threshold. The peak of this function yields the optimal AO, though it is often worth

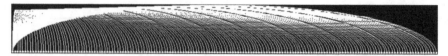

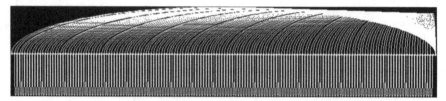

Fig. 6.4 Audio Oracle structure built on Prokofiev work – distance threshold was chosen to be 0.000001, and total CR was 0

Fig. 6.5 Audio Oracle structure built on Prokofiev work – distance threshold was chosen to be 0.9, and total CR was 0

studying oracles formed using secondary peaks as well. Figure 6.4 shows an oracle where the distance threshold was too small. In this case, no similarity between any sound frames was found, and all frames were considered to be new. Each frame has only a link from state 0, indicating that this is its first appearance in the sequence. AO has determined the input to be a "random" sequence, where all frames are unrelated and no repetitions or patterns occur. Conversely, Fig. 6.5 shows an oracle where the distance threshold was set too high, lumping disparate frames together, and producing an oracle where all frames were considered to be repetitions of the first. Each frame has a suffix link to the previous frame, indicating this repetition. Note that in these diagrams, suffix links to the 0th frame, present when a new frame is detected, are omitted for visual clarity. In the case of the Prokofiev work, the optimal distance threshold was found to be 0.21, which yielded a total CR of 1592.7.

6.11 Analyzing "Visions Fugitives"

In order to experiment with CR measure we analyzed a solo piano work by the composer Sergei Prokofiev. We chose the fourth movement of the "Visions Fugitives" (1917) because of its short duration and clear form. The piece is in a simple binary form, with a highly active and dynamically varied A section contrasting with a more repetitive and soft B section. The A section, mm. 1–28, consists of a series of short thematic materials, and has a periodic four bar phrase length throughout. Section A is subdivided into 6 subphrases, built using three main materials. The first two phrases, in mm. 1–4 and 5–8, present materials A1 (shown in Fig. 6.6) and A2 (Fig. 6.7). A1 is characterized by a descending pat-tern built around a half-step. A2 builds on this half-step idea, but with an emphasis on harmony. The introductory material, A1, returns in modified form as A1' in mm. 9–12 and A1" in mm. 13–16. New material,

Fig. 6.6 *A1* Material from Prokofiev work

Fig. 6.7 *A2* Material from Prokofiev work

Fig. 6.8 *A3* Material from Prokofiev work

based on a descending melodic pattern played over descending arpeggiated major 7th chords, is introduced in m. 17. This material, labeled A3 (Fig. 6.8), continues until m. 21, where a modified version of A2 is played. This A2' combines the harmonic and rhythmic character of A2 with the descending melodic pattern of A3. Four bars of transitional material, labeled T, follow from mm. 25–28, and connect the half-step motion of section A with the ostinati built on thirds of section B. The B section, from mm. 29–49, consists of three longer sub-phrases of uneven duration. The first, running from mm. 29–33, is labeled B1. This phrase establishes the harmonic context of the section, and combines the half-step patterns from before with a minor third ostinato. B2 and B2', mm. 34–39 and 40–49 respectively (Fig. 6.9), embellish this material with a simple half-step melody. The piece ends with a single, sustained low G. The clear form of the piece allows for better appreciation of the role of expectation and explanation in forming the musical structure. Figure 6.10 shows our analysis of the work in terms of sectionality and related phrases. It is important to note that there is a "semantic gap" between the signal features used in CR analysis and our subjective musical analysis. We do not claim that CR measures

Fig. 6.9 *B2'* Material from Prokofiev work

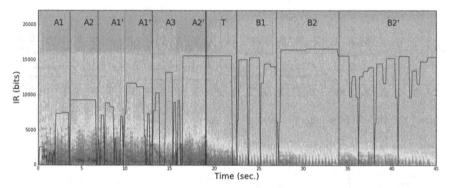

Fig. 6.10 Formal structure of Prokofiev 'Visions Fugitives Mvt. 4'—Information Rate is plotted against the signal spectrogram

or recognizes concepts like harmony, melodic contour, or rhythmic motives, but it seems apparent that low-level signal features to relate to these concepts in some way. Both our sectional analysis and the CR measure are shown on top of the signal spectrogram. The audio features used for analysis were 38 cepstral coefficients estimated over signal frames of 8192 samples (approximately 186 ms) in duration. Of particular note is the way in which the CR measure corresponds with our subjective musical analysis of the work. CR shows an overall pattern of growth over the piece, as the ability to explain new materials as repetitions of past ones increases. As new materials are introduced, CR momentarily drops, and many of the significant drops in CR correspond to section or phrase boundaries. As an example, consider sections A1, A1', and A1". A1 is the first material of the piece, so its CR is low. As the phrase passes, it becomes possible to explain the new sound frames as being repetitions of past frames. Each successive variation on A1 exhibits higher CR than the previous, and it is particularly revealing to examine the CR contour of A1' and A1". These variations share a similar shape, with the CR of A1" slightly higher than that of A1'. As we reach the second half of the piece, the B section, CR blocks become longer in length, and moments of low CR become shorter. This is due to two factors: first, the material during the second section is more repetitive than that in the first, and second, an overall larger number of frames can be explained by the past.

The Audio Oracle used to perform this CR analysis is the same as that pictured in Fig. 6.3. It is possible to observe aspects of the musical structure directly from

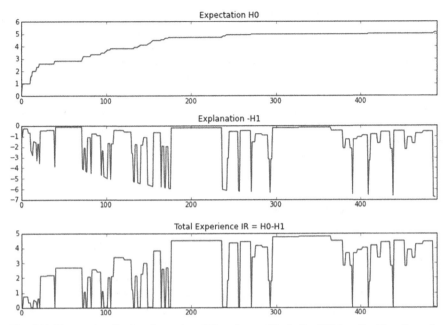

Fig. 6.11 Expectation, Explanation and total Experience of Prokofiev "Visions Fugitives," as calculated from entropies of encoding the MFCC features without and with considering the repetition structure as found by the Audio Oracle

the AO graph. For example, the dynamic and constantly changing section A corresponds to the many forward links from the first sound frame. Each of these forward links corresponds to the appearance of a new frame. In contrast, section B consists of repetitions of a few sound frames in varying patterns. This is in agreement with our structural analysis of the work as well.

The graph in Fig. 6.11 shows the three elements H_0, H_1 and CR over time for a recording of the Prokofiev. The top graph shows the increase in the coding length as more new frames appear over the time course of the composition. This graph follows the general shape of a logarithmic curve, with flat segments corresponding to regions where the existing alphabet is used to represent the musical events, and vertical jumps occurring every time a new and yet unheard frame is added to the musical materials. The second frame represents the Explanation part, captured by negative entropy computed using the AO compression method. Starting from high Explanation during the first appearance of the A1 theme (first materials are encoded/ explained as repetition of a single first event), the Explanation drops as more new sounds appear that cannot be traced back to previously heard materials. As time passes, more repetitions are detected and thus longer segments in the song can be "explained" in terms of recopying of earlier sounds. As seen in the final graph and Fig. 6.10, drops in explanation are often related to transitions between sections of the song. The overall experience is considered as a sum of expectation and explanation, which in our case is taken with equal weights. The question of weighting the expectation and explanation components is left for future research.

It should be also noted that our analysis depends on the similarity threshold that is used by AO to detect the repetitions. High thresholds tend to capture longer repetitions, but they adversely affect the expectation graph since fewer new states are detected. Low thresholds tend to favor higher expectation (more new materials are detected) but this adversely affects the ability to find repeating segments, thus lowering the explanation graph. In the graphs shown in Fig. 6.11, we used a difference threshold that was found by taking the maximal total CR value (area under the CR graph) resulting from our analysis. For an evaluation of the AO- CR analysis with a more detailed analysis of Beethoven Piano Sonata No. 1 in F minor, the reader is referred to (Dubnov et al. 2011).

6.11.1 Creative Applications

One of the practical results of our approach is developing a firm methodology as well as tools so that AO can be used as a generator (high level instrument) in computer aided composition. For the on-line, real-time interaction approach, two main systems have been built: OMax by Assayag, Bloch and Chemillier, an OpenMusic/Max application for co-improvisation with human performers using MIDI, audio and/or video media; and MiMi by A. François and E. Chew, a multi-modal interface for co-improvisation with the computer that gives visual feedback to the performer, helping him to anticipate the computers reactions. A new OMax variant, called So-Max, is currently being developed. SoMax analyzes a corpus of multiple works, and uses this data to generate accompaniment to a live performance. OMax operation is usually done as a trio—human musician-improviser, machine improviser and human operator of the machine. The role of the humans in this setting is quite demanding—the musician creates initial musical materials and then listens and responds to machine improvisation. The human operator listens to the human musician and controls the machine using his own taste and musical skills.

Both OMax and Mimi use symbolic Factor Oracle structures, using MIDI input or quantizing input audio to a discrete alphabet. OMax also comes with presets of specific features and settings that were manually optimized for specific instruments. An advantage of AO over this approach is the automatic method for finding the best oracle, described above. In addition, the ability to use oracles built on multiple features provides a greater level of flexibility. Through CR, it is possible to determine which feature provides the most information at a given time, and choose to focus oracle improvisation on that feature.

6.11.2 Meta-Creation

In this process of using the machine for creating an interesting/aesthetic musical form, we separate between the steps of providing musical input to the machine and control of machine improvisation. The recording, analysis and design of the final musical results can be done in real time or off-line. In our early experiment, the design of the work is done by running the oracle according to a compositional plan that has been automatically created accordingly to our aesthetic reward model.

The process of creating the composition consists of the following steps:

- Performing AO analysis using corpus of pre-recorded music materials of certain style
- Performing Information Dynamic analysis of the AO graph structures for multiple features and music parameters
- Derivation of reward functions from the Information Dynamics signal
- Creating the composition design according to computer suggestions for optimal reward trajectory (policy).
- Generating a musical score and performing the new piece

In (Dubnov et al. 2011) we showed how the sum of CR values over the whole signal can be used to find the optimal threshold for the AO estimate. Generalizing this idea, we consider CR values for different features by analyzing the structure of AOs constructed according to different acoustic parameters. In practice, the most common features used for improvisation are pitch and spectral features, providing two oracles that give different structures which the improvisation follows. In musical terms, the spectral oracle allows for timbral improvisation, or in other words, structuring of the composition according to changes in sound color, while ignoring things such as harmony or melody. The pitch oracle captures aspects of music that are governed by its melodic structure. Accordingly, CR can be used not only for creating variations by recombination of segments within an oracle, but also for switching between "most informative" or "most aesthetic" oracles at different points in time, a phenomena that could be considered as selective attention.

In addition to OMax and MiMi, a third system has been developed. PyOracle, by Surges and Dubnov, provides a free software toolkit for experimentation with oracle structure and CR measurement, and can also be embedded in a real-time computer music environment like Max/MSP or Pure Data (Surges 2013). Unlike the Factor Oracle-based systems described above, PyOracle uses Audio Oracle. This means that PyOracle does not quantize input sounds to a discrete alphabet and can be calibrated through CR. In a live improvisation, a human musician improvises, providing musical materials and structure. The audio of this performance is both recorded to a buffer and subjected to feature-analysis. Arbitrary features can be tracked, allowing for spectral, pitch, or some other type of oracle to be built. Each feature frame is given a timestamp, corresponding to its appearance in the audio buffer. An AO structure is built, one frame at a time, as new sound frames enter the system. The online nature of the AO construction algorithm makes this relatively undemanding computationally.

During improvisation, the machine improviser navigates through the oracle. At each oracle state, the improviser moves linearly forward with probability p or jumps forward or backward along an oracle suffix link with probability $(1-p)$. This probability p is exposed as a parameter to the machine operator, and provides a high-level control over the ratio of the original material to newly generated recombinations. Recalling that since oracle positions connected by suffix links share some amount of common context, these jumps should be relatively perceptually smooth. The oracle *LRS*, as defined above, is also exposed to the operator. In the context of oracle improvisation, *LRS* is another useful parameter. Since the *LRS* refers to the length of the common suffix between two positions in the audio signal, we can consider it as a measure of shared musical context. The amount of shared context between two positions influences the perceived smoothness of a jump between them. We can use *LRS* as a threshold on possible suffix link jumps enforcing that jumps only occur between positions with a specified amount of shared context.

In the most general case, the machine improviser has access to the entire range of the signal. However, a level of compositional control can be attained by restricting the machine improviser to only a specific time region of the signal and its oracle representation. If a region is set, the oracle pointer will be constrained to only navigate within that region. If the region is defined with care, this is akin to asking a human performer to improvise on only a specific type of material. Regions can be specified in real-time by the machine operator or according to a precomposed score file. Use of regions in real-time improvisation allows the machine operator to enter into a dialog with the human improviser by recalling previous materials according to his or her taste and the development of the improvisation. Score files, which are described below, allow for the pre-composition of specific musical structures which retain a sense of identity across performances, though the actual musical contents may vary widely.

We have explored three strategies for composition or structured improvisation using PyOracle:

- Using a pre-composed score file.
- Duo performance with human machine operator.
- Solo performance with machine-defined segmentation.

The first of these strategies, using a pre-composed score file, is perhaps the most rigid. An improvisation "script" is defined, with timed or un-timed instructions for the PyOracle software. Any of the PyOracle parameters can be controlled by the script file, but some of the most useful are: output gain, region selection, *LRS* threshold, and recombination probability. In the simplest case, each section of a work is defined according to a set of parameters, and the sections are stepped through linearly according to pre-defined durations. This allows for arbitrarily complex formal structures, but, as mentioned, the specificity of this strategy comes with a certain amount of formal rigidity, which may be somewhat constraining to an improviser. As a compromise between formal specificity and improvisational freedom, it is possible to use a MIDI foot pedal or other controller to advance the script, instead of the machine clock. It might also be desirable to build some kind of 'if-then-else' logic into the script at certain points. For example, at the end of a section, the software could be instructed to branch to one of multiple possible continuations, based on what the instrumentalist plays.

To enable the emergence of spontaneous formal structures, it may be desirable for a second human performer to operate the machine. The machine operator is then

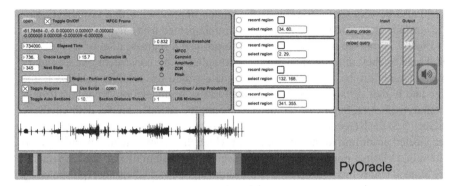

Fig. 6.12 Real time interactive use of Audio Oracle to create new improvisations on incoming audio. Lower colored bars indicate real-time CR segmentation

responsible for defining specific regions of the improvised performance which he can then recall at-will. By constraining the Oracle navigation to a specific region, defined according to whatever musical parameters are deemed relevant, the machine operator is able to create agreement or contrast with the instrumental improviser. PyOracle has a feature which enables rapid selection and recall of regions. The start- and end-points of a region can be selected with a simple toggle button, and the captured region can be recalled with the click of another button. The interface allows for multiple regions to be stored and overwritten as desired as the performance develops. This control method allows for rapid and spontaneous identification, capture, and re-use of musical materials generated by the instrumentalist.

Figure 6.12 shows the PyOracle software. PyOracle is written in a combination of the Python programming language and Max/MSP, a real-time computer music programming environment, which allows musicians to program audio and multimedia software using "objects" connected by "patch cords". The interface is divided into four main sections. The upper left section provides access to Oracle navigation parameters, like the jump probability and regions. It also allows the user to adjust the distance thresholds for AO and SO construction. To the right is a bank of region selection controls, as described above. The last panel in the upper row provides output gain and debugging controls. The bottom row presents a waveform view of the input signal, which is overlaid with a representation of the Oracle navigation pointer (vertical bar) and the currently defined region (shaded region).

Due to the modular construction of PyOracle, and the high-level nature of both Python and Max/MSP, the system is easy to extend. Enhancements and modifications to the real-time improvisation system can be easily constructed on an ad-hoc basis, as particular musical or experimental needs dictate. Future improvements and enhancements include audio-driven oracle navigation where input audio constrains the oracle navigation to specific regions or materials, "selective listening" using multiple oracles constructed simultaneously on different features, and more facilities for structural organization of improvisations. We also plan to further explore the use of PyOracle in guided improvisation and open-form compositions. It could also be very useful to provide the human improviser with a visualization of the machine state, and this could be easily implemented using PyOracle.

Finally, it may be possible to use CR to provide a segmentation of the musical materials as they unfold. This can be accomplished by constructing a secondary Audio Oracle on the input audio. This Oracle would use a larger analysis frame size and higher distance threshold, to avoid false segmentation according to momentary contrasts. Segments are detected using the encoded blocks defined above in Sect. 9.3. Each (*length, position*) pair corresponds to one segment, and seg-ments are labeled according to the position from which they copy. An automatic segmentation of a clarinet improvisation can be seen at the bottom of Fig. 6.12. Segments are aligned below their occurrence in the audio signal, and labels are represented by color. If a segment shares a label, and therefore a color, with an earlier segment, this represents a repeated section.

PyOracle could navigate this segmentation Oracle (SO) in a similar way to the navigation algorithm described above. During an improvisation, the SO is traversed either linearly or via jumps, according to a user-specified probability. If a jump is chosen, the SO first attempts to jump to a segment with the same label—a repeated section. If no similarly labeled segment exists, then a random segment is chosen. In both cases, longer segments are favored.

6.12   Conclusion: The Challenge of Composing and the Pleasure of Listening

In this paper we presented a novel framework for modeling music that is based on interaction between two apparently conflicting music behaviors—composition that tries to generate novel auditory experiences, and listening that tries to make sense of these sensations. We assume that the task of the composer is to maximize, within certain stylistic constraints, the amount of information that the music delivers to the listener. The listener, on the other hand, tries to maximize his understanding by mapping the incoming stimuli to his pre-existing knowledge and memory. These trends are contradictory in the following sense—the more variations and changes are injected by the composer, the more difficult it is going to be for the listener to recognize and capture the structure of the incoming music. On the other hand, maximizing the listener understanding will result in music that is perfectly matching to existing schemes and memory, creating a rather unappealing experience. Accordingly, some optimal balance should exist between these two trends. Also one should note that these two modalities can exist within a single person—a composer is also a listener, and a listener also has a drive for excitement and experience seeking behavior.

The computational modeling framework that informed our approach is based on a combination of several lines of research—work on Music Information Dynamics that characterizes changes in information contents of audio signals, work on automatic detection of music repetition structure using Audio Oracle, and research on decision making for finding aesthetically optimal policy in traversing the graph of musical possibilities during the music generation phase.

Accordingly, we suggested a three tiered model: a musical memory model that is capable of detecting and storing references to previously heard musical materials, a listening appraisal model that evaluates how well new materials can be "explained" in terms of references to existing memory, and a generative model that composes new materials by recombinations of musical materials stored in the AO memory. In other words, the process of musical composition is viewed here in terms of an action-cognition loop that uses Information Dynamics as a signal for navigating the Audio Oracle graph. By using Information Dynamics as an indicator for the success or failure of higher level mental listening faculties, one can asses in computational manner the so-called "cognitive mastering" aspects of aesthetic perception. Using Audio Oracle allows modeling of musical structure by learning directly from a corpus of musical works, surpassing the need for expert encoding of musical knowledge of every specific style or genre. The Audio Oracle has been successfully used in machine improvisation and computer aided composition applications. Moreover, listening aspects such as emotional force, familiarity and other cognitive attributes in music perception were successfully modeled in terms of changes in information contents of musical signals using Information Dynamics.

References

Abdallah S, Plumbley M (2009) Information dynamics: patterns of expectation and surprise in the perception of music. Connect Sci 21(2–3):89–117

Allauzen C, Crochemore M, Raffinot M (1999) Factor oracle: a new structure for pattern matching. SOFSEM99: theory and practice of informatics, pp 295–310

Baggi DL, Haus G (2009) The new standard IEEE 1599, introduction and examples. J Multimedia 4(1):3–8

Bense M (1969) Introduction to the information-theoretical aesthetics, foundation and application to the text theory. Rowohlt Taschenbuch Verlag (in German)

Bewley J (2004) Lejaren A. Hiller: computer music pioneer. Music Library Exhibit, University of Buffalo, 2004. http://www.cse.buffalo.edu/rapaport/111F04/hillerexhibitsummary.pdf

Birkhoff GD (1933) Aesthetic Measure. Cambridge, Mass

Burns K (2006a) Atoms of EVE': a bayesian basis for esthetic analysis of style in sketching. AI EDAM 20:185–199

Burns K (2006b) Fun in Slots. Proceedings of the IEEE conference on computational intelligence in Games. Reno

Chomsky N (1957) Syntactic Structures. Mouton, The Hague

Collins K (2009) An introduction to procedural music in video games. Contemp Music Rev 28(1):5–15

Conklin D (2003) Music generation from statistical models. Proceedings of symposium on AI and creativity in the Arts and Sciences, pp 30–45

Dubnov S (2006) Spectral anticipations. Comput Music J 30(2):63–83

Dubnov S (2008) Unified view of prediction and repetition Structure in audio signals with application to interest point detection. IEEE Trans Audio, Speech Lang Proc 16(2):327–337

Dubnov S, Assayag G (2002) Universal prediction applied to music genertion with style. Mathematics and Music, Springer

Dubnov S, Assayag G, Lartillot O, Bejerano G (2003, Oct) Using machine-learning methods for musical style modeling. IEEE Comput 36(10):73–80

Dubnov S, McAdams S, Reynolds R (2006) Structural and affective aspects of music from statistical audio signal analysis. J Am Soc Inf Sci Technol 57(11):1526–1536

Dubnov S, Assayag G, Cont A (2007) Audio oracle: a new algorithm for fast learning of audio structures. Proceedings of the International Computer Music Conference (ICMC)

Dubnov S, Assayag G, Cont A (2011) Audio oracle analysis of musical information rate. The Fifth IEEE International Conference on Semantic Computing, Palo Alto

Electronic Arts (2013) What is spore? Spore.com. http://www.spore.com/what/spore. Accessed 3 July 2013

Farnell A (2007) An introduction to procedural audio and its application in computer games. Audio Mostly Conference

Gallez DW (1976) Satie's "Entr'acte:" a model of film music. Cinema J 16(1):36–50

Huron D (2006) Sweet anticipation: music and the psychology of expectation. MIT Press

Itti L, Baldi P (2005) Bayesian surprise attracts human attention. Adv Neural Inform Process Syst 19:18

Koster RA (2005) Theory of fun for game design. Paraglyph Press, Scottsdale

Kundera M (1981) Prophte de l'Insensibilit. In: Fleuret M (ed) Regards Sur Iannis Xenakis

Leder H, Belke B, Oeberst A, Augustin D (2004) A model of aesthetic appreciation and aesthetic judgments. Brit J Psychol 95(4):489–508

Lefebvre A, Lecroq T (2001) Compror: compression with a factor oracle. Proceedings of the Data Compression Conference

Krull R, Watt JH Jr, Lichty LW (1977) Entropy and structure: two measures of complexity in television programs. Commun Res 4:61

Lewis GE (2000) Too many notes: computers, complexity, and culture in voyager. Leonardo Music J 10:33–39

Meyer LB (1956) Emotion and meaning in music. Chicago University Press, Chicago

Pequeno Ze (2013) Autechre: interview. Tiny mix tapes. http://www.tinymixtapes.com/features/autechre. Accessed 14 April 2013

Potter K, Wiggins GA, Pearce MT (2007) Towards greater objectivity in music theory: information-dynamic analysis of minimalist music. Musicae Scientiae 11(2):295–322

Potter RF, Choi J (2006) The effects of auditory structural complexity on attitudes, attention, arousal, and memory. Media Psychol 8(4):395–419

Pritchett J (1996) The music of John Cage. Cambridge University Press, Cambridge

Prokofiev S (1955) Visions fugitives. 1915–1917. In: Prokofiev S (ed) Collected works, vol. 1. Moscow, Muzgiz

Rigau J, Feixas M, Sbert M (2008, March–April) Informational aesthetics measures. IEEE Comput Graph 28(2):24–34

Salimpoor VN, Benovoy M, Larcher K, Dagher A, Zatorre RJ (2011) Anatomically distinct dopamine release during anticipation and experience of peak emotion to music. Nat Neurosci 14(2):257–264

Schultz W (2006) Behavioral theories and the neurophysiology of reward. Annu Rev Psychol 57:87–115

Spehara B, Cliffordb CWG, Newellc BR, Taylor RP (2003) Universal aesthetic of fractals. Comput Graph 27:813820

Surges G (2013) Pyoracle. https://bitbucket.org/pucktronix/pyoracle. Accessed 14 April 2013

Weiner M (2013) Brian Eno and the Ambient Series, 1978–1982. Stylus Magazine. http://www.stylusmagazine.com/articles/weeklyarticle/brian-eno-and-the-ambient-series.htm. Accessed 14 April 2013

Ziv J, Lempel A (1977) A universal algorithm for sequential data compression. IEEE T Inform Theory 23(3):337–343

Ziv J, Lempel A (1978) Compression of individual sequences via variable-rate coding. IEEE T Inform Theory 24(5):530536

Chapter 7
Machine Listening of Music

Juan Pablo Bello

7.1 Introduction

The analysis and recognition of sounds in complex auditory scenes is a fundamental step towards context-awareness in machines, and thus an enabling technology for applications across multiple domains including robotics, human-computer interaction, surveillance and bioacoustics. In the realm of music, endowing computers with listening and analytical skills can aid the organization and study of large music collections, the creation of music recommendation services and personalized radio streams, the automation of tasks in the recording studio or the development of interactive music systems for performance and composition.

In this chapter, we survey common techniques for the automatic recognition of timbral, rhythmic and tonal information from recorded music, and for characterizing the similarities that exist between musical pieces. We explore the assumptions behind these methods and their inherent limitations, and conclude by discussing how current trends in machine learning and signal processing research can shape future developments in the field of machine listening.

7.1.1 Standard Approach

Most machine listening approaches follow a standard two-tier architecture for the analysis of audio signals. The first stage is devoted to extracting distinctive attributes, or *features*, of the audio signal to highlight the music information of importance to the analysis. The second stage utilizes these features either to categorize signals into one of a predefined set of classes, or to measure their (dis)similarity to others.

In the literature, the first stage typically utilizes a mix of signal processing techniques with the heuristics necessary to extract domain-specific information.

J. P. Bello (✉)
Music and Audio Research Laboratory (MARL),
New York University, New York, USA
e-mail: jpbello@nyu.edu

N. Lee (ed.), *Digital Da Vinci*, DOI 10.1007/978-1-4939-0536-2_7,
© Springer Science+Business Media New York 2014

Unsurprisingly, feature extraction methods tend to be specific to music, or shared with closely-related problems in speech processing or the analysis of environmental sounds. The second stage tends to rely on multi-purpose pattern matching or recognition methods that exploit statistical regularities in a domain-agnostic way. Therefore, it is not rare to find the same classification, regression and clustering techniques in music informatics literature that are found in computer vision, bioinformatics and machine learning work.

From the above, it is clear that the uniqueness of machine listening work in music is derived from the particularities of the feature extraction process. But, why extracting features in the first place? From an information processing perspective, attempting to recognize patterns directly on raw, audio data is simply impractical. There are nearly 90,000 real values contained in one second of CD-quality audio, and more than 200 s to the average pop song. Most of these values represent information that is too noisy, repetitive or irrelevant for music analysis. The process of feature extraction projects the audio signal into a smaller, more informative space where robust analysis is feasible. This space is typically multi-dimensional since several features are needed to appropriately describe the music contents of the signal.

Interestingly, early efforts at designing more and better features for automatic music analysis, have been progressively replaced by the use of ever more-powerful pattern recognizers on a dwindling set of standard features. This trend is underpinned by widely-reported increases in recognition accuracy resulting from increases in complexity, and thus the modeling power, of these techniques, and is sustained by the growing adoption of feature extraction toolboxes and pre-computed feature sets, which entrench current features as the best, or only, solution to entire families of problems in the automatic analysis of music.

However recent research, including our own work on the systematic evaluation of chord recognition techniques, demonstrates that the impact that feature design and low-level processing have on performance significantly outweighs the benefits of using complex pattern matching strategies (Cho et al. 2010; Cho and Bello 2014). Further, that a nuanced understanding of the limitations of current features and the interplay between the different processing stages, is necessary for the development of robust machine listening approaches.

The following sections will introduce the reader to the most common techniques used for the analysis of timbre, rhythm and tonal information in music with a focus on feature extraction methodologies. This chapter aims to provide some intuition for current methods, and identify limiting assumptions that should be taken into account when implementing machine listening systems.

7.2 Timbre

For how important timbre is to the creation, perception and analysis of music, it is remarkable how difficult it is to agree on a definition. Timbre guidelines in musical scores and digital representations such as MIDI, are often limited to the assignment of instrumental classes, and rarely address or control the specific set of qualities that define a particular sound or sound source. This is largely because it is unclear what those qualities are, and how to describe them. Thus, unsurprisingly, much of the literature defines timbre in terms of what it is not, i.e. the attribute that allows us to differentiate sounds of the same pitch, loudness, duration and spatial location.

Following the pioneering work of (Grey 1975; Wessel 1979), much progress has been made on identifying perceptual qualities that are relevant to timbre. This is done by empirically measuring the perceived similarity between sounds and using those measurements to project to a low-dimensional timbre space, where dimensions are assigned semantic interpretations such as brightness, temporal variation or synchronicity. From an audio perspective, these timbre spaces can be recreated using low-level acoustic features with similar interpretations, such as the spectral centroid, spectral flatness, attack time, etc. Interestingly, many of these features provide descriptions of the *spectral envelope* of the sound, and were originally proposed in the context of speech analysis.

7.2.1 Spectral Envelope

Simply put, the human speech system can be thought of as a subtractive synthesizer where a source signal, the oscillations produced by the vocal folds, are filtered by the combined action of the tongue, lips, jaw, velum and larynx and the physiology of the nose and mouth cavities (Cook 2001). This source-filter view has long been adopted in speech processing for the modeling of audio signals such that, in the frequency domain, the source component accounts for the fast-changing spectral line structure most often associated with pitch, while the filter component accounts for the slow-changing spectral envelope, or shape, which is most often associated with timbre.

The coarsest approximation of the spectral envelope is as a simple, uni-modal distribution defined by two parameters: the spectral centroid and the spectral spread. As the name indicates, the spectral centroid can be conceptualized as the center of gravity of a magnitude spectrum $|X(m, k)|$, and defined as:

$$SC(m) = \frac{\sum_k f_k |X(m,k)|}{\sum_k |X(m,k)|} \tag{7.1}$$

where m, k are, respectively, the time and frequency indexes of the spectral analysis, and f_k is the frequency, in Hertz, of the kth spectral bin. The centroid is usually associated with the sound's brightness. Likewise, the spectral spread can be defined as:

$$SS(m) = \frac{\sum_k (f_k - SC(m))^2 |X(m,k)|}{\sum_k |X(m,k)|} \tag{7.2}$$

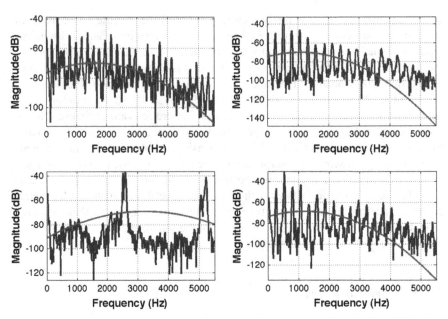

Fig. 7.1 Spectral envelope approximation using centroid and spread. The *blue line* depicts the magnitude spectrum, while the *red line* depicts the approximation. The *horizontal axis* represents frequency in Hertz. The *vertical axis* represents magnitude in dB

and understood as a measure of the bandwidth of the spectrum. Together, spectral centroid and spread can be used to fit a normal distribution to the spectrum, resulting in the coarse, red-line approximation observed in Fig. 7.1. The advantage is, of course, that the thousands of values required to represent the spectrum of a complex sound have been reduced to only two features.

A more accurate approximation can be obtained by means of the *channel vocoder* approach, which decomposes the sound using a bank of bandpass filters, and then summarizes the magnitude for each bandpass signal. For a set of L-long filters w, overlapped by $L-1$ bins, the channel-vocoder operation can be seen as a circular convolution of the form:

$$\text{CV}(m) = |X(m,k)| * w(k) \tag{7.3}$$

Thanks to the convolution theorem (Smith 2007), this operation can be efficiently implemented as the product of fast Fourier transforms:

$$\text{CV}(m) = \mathbb{R}(\text{IFFT}\,[\text{FFT}\,(|X(m,k)|) \times \text{FFT}\,(w(k))]) \tag{7.4}$$

where $w(k)$ is normalized to unit sum, zero-padded to the length of X, and circularly-shifted such that its center coincides with the first bin. The characteristics of the resulting spectral envelope approximation, the red line in Fig. 7.2, depends on the

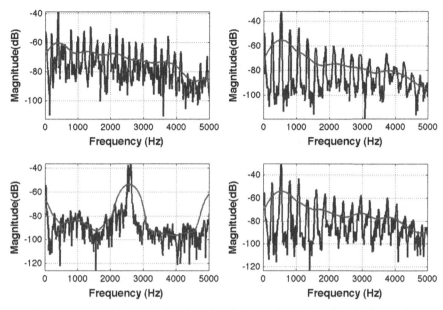

Fig. 7.2 Spectral envelope approximation using the channel vocoder. The *blue line* depicts the magnitude spectrum, while the *red line* depicts the approximation. The *horizontal axis* represents frequency in Hertz. The *vertical axis* represents magnitude in dB

choice of window length L. A low-dimensional feature vector, typically with tens of coefficients, can be obtained simply by downsampling this curve.

7.2.2 Cepstrum and MFCC

Another powerful strategy is based on the idea of operating in the so-called "cepstral" domain. The main idea behind cepstral analysis is to take a log-magnitude spectrum and treat it as if it were a time-domain signal, such that the rate of change across frequency bands is measured by taking the (inverse) discrete Fourier transform (Oppenheim and Schafer 2004). Since the transform is done on the log-magnitude spectrum, the resulting representation is not in time, but on an alternate inverse spectral domain, the cepstrum, which is a function of lag. For a real-valued signal x, this is defined as:

$$c_x(l) = real(IFFT(log(|FFT(x)|)))$$ (7.5)

Just as with time-domain signals, slow components such as the spectral envelope are encoded on the low-lag coefficients of the cepstrum. Therefore, we can weigh the real cepstrum using a low-pass window that zeroes all components above a chosen cutoff lag L_1. The resulting spectral envelope approximation, the red-line in Fig. 7.3, can be visualized by transforming back the low-pass filtered cepstrum c_{LP}:

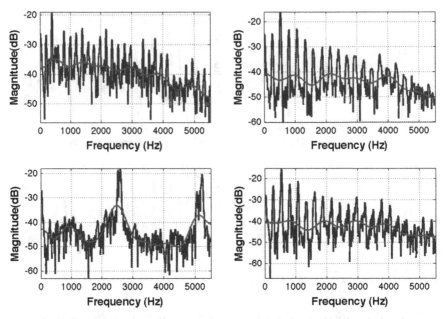

Fig. 7.3 Spectral envelope approximation using cepstral analysis. The *blue line* depicts the magnitude spectrum, while the *red line* depicts the approximation. The *horizontal axis* represents frequency in Hertz. The *vertical axis* represents magnitude in dB

$$C_{LP}(k) = e^{\Re\left[\text{FFT}(c_{LP}(l))\right]} \tag{7.6}$$

Notably, only a few coefficients in the cepstral domain ($L_1 = 15$ in this example) are sufficient to robustly approximate the spectral envelope.

The Mel-Frequency Cepstral Coefficients (MFCCs), a variation of the linear cepstrum coefficients presented above, is by a considerable margin the most widely-used feature representation in machine listening research. MFCCs include two simple but crucial differences in their computation. First, instead of using linearly-spaced frequency analysis, we use the *Mel* scale, a non-linear perceptual scale of pitches judged to be equidistant. The relationship between mel and frequency in Hertz is defined by;

$$\text{mel} = 1127.01028 \times \log\left(1 + \frac{f}{700}\right) \tag{7.7}$$

which results on a scale that is approximately linear below 1 kHz, and logarithmic above that. The reference point is at $f = 1$ kHz, which corresponds to 1,000 mel, such that a tone perceived to be half as high is 500 mel, one twice as high is 2,000 mel, etc. The warping from linear to mel frequency is typically performed by means of a filterbank of half-overlapping windows, with center frequencies uniformly distributed in the mel scale.

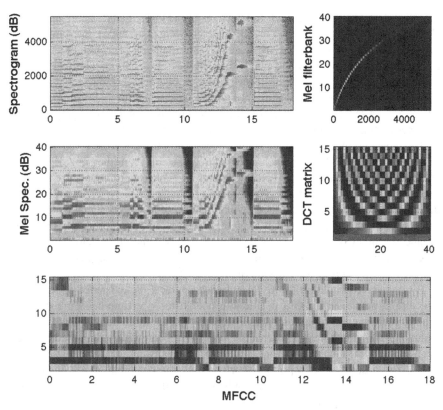

Fig. 7.4 Computation of MFCCs for a violin signal: the signal's spectrogram (*top left*) is passed through a bank of bandpass filters with center frequencies organized according to the Mel scale (*top right*). The resulting Mel spectrogram (*middle left*) is transformed using the DCT (*middle right*), to obtain the sequence of MFCCs (*bottom*)

The second variation in MFCCs is to use a discrete cosine transform (DCT), instead of the DFT, to obtain the cepstral representation. This exploits the compression properties of the DCT (deriving from the symmetry of the DCT reconstruction, in turn a by-product of the evenness of cosines), to project most of the information in the spectrum into only a few, approximately decorrelated Mel-frequency cepstral coefficients. The entire process is illustrated in Fig. 7.4. For visualization purposes, we only show coefficients 2–15, since the magnitude values of the first two coefficients (closely correlated to measures of spectral energy and slope) are much higher than the rest.

7.2.3 Sound Similarity and Classification

MFCCs and other features typically describing the spectral envelope of sounds, constitute the basis of most sound similarity and classification approaches in music, speech and environmental sound. In music, example tasks include genre and artist classification (Tzanetakis and Cook 2002; Berenzweig et al. 2003), instrument recognition (Herrera et al. 2006; Barbedo 2012), auto-tagging (Turnbull et al. 2008), automatic recommendation and playlisting (McFee et al. 2012). The majority of these applications operate at the level of sounds or tracks lasting seconds if not minutes. On the other hand, features are typically extracted for audio snippets only a fraction of a second long. Therefore it is necessary to combine, or aggregate, features across time into sound- or track-level feature representations.

The most common method for feature aggregation through time is the so-called "bag of features" (BOF) approach. This approach stems from the "bag of words" assumption for the analysis of text documents, in which word counts are used to represent documents without regard to phrase structure. Likewise, sounds can be described in terms of the general behavioral pattern of low-level signal features regardless of the specific values they assume, or their temporal ordering and structure (Tzanetakis and Cook 2002; Aucouturier et al. 2007). This can be achieved simply by constructing a feature histogram or by modeling the feature sequence either as one or a mixture of multivariate Gaussian distributions.

By its very formulation the BOF approach neglects the existence of temporal structure in music, and is thus unable to properly describe time-based musical attributes such as melody, harmony or rhythm. In fact, the limitations of this strategy, including the symptomatic existence of *hubs*, have been widely documented (e.g. in (Aucouturier 2006; Aucouturier et al. 2007; Berenzweig 2007; Schnitzer et al. 2012). Yet, despite these important shortcomings, the combination of MFCC-like features and the BOF approach has resulted in remarkable success for a variety of music informatics applications, including systems that are already available in the market (The Echonest 2013; BMAT 2013; Gracenote 2013).

7.3   Rhythm

Variously defined as "organized duration" (Kolinski 1973), "the many different ways in which time is organized in music" (Bamberger and Hernandez 2000), and the "language of time" (Lewis 2007), the attribute of rhythm is of central concern to most music analyses. Nevertheless, it is only in recent decades that the subject has received the attention it deserves, producing rhythmic theories on a par with those for pitch structure; musicological studies providing a more nuanced understanding of the historical and cultural context that informs the creation and reception of rhythm; as well as new science allowing us to understand the biological mechanisms behind rhythm processing, and the effects of rhythm information on human perception and cognition (London 2012; Agawu 2012; Janata et al. 2012; Honing 2012; Toussaint 2013).

In machine listening research, rhythm-based work is typically under-pinned by the tasks of identifying the start time, or *onset*, of new events in the music signal, a problem known as *novelty detection*; and the task of characterizing the patterns of (quasi-)periodicity and accentuation that define beat and metrical-level structure in the music. The following discusses these analysis steps in more detail.

7.3.1   *Novelty Detection*

Onsets are single instants in time chosen to mark the start of the (attack) transient of a new event. Transients are transitional regions of the signal, characterized by their short duration, instability and, in the case of attack transients, a sudden burst of signal energy (Bello et al. 2005). They are also often followed by the so-called steady-state of the sound, where content is much more stable and predictable.

Detecting onsets can be used for signal segmentation (Jehan 2005), to inform tempo and beat tracking approaches (Davies and Plumbley 2007), to analyze expressive performance practice (Widmer et al. 2003), and for sound manipulation and synthesis (Ravelli et al. 2007; Hockman et al. 2008), to name only a few examples. Yet, the detection process can be affected by a number of difficulties, including the existence of extended transients (e.g. in bowed strings) which cannot be appropriately characterized by a single time instant; the presence of ambiguous events such as vibrato, tremolo or glissandi, where the distinction between existing and new events is not clearly defined; and the analysis of polyphonic and/or multi-instrumental ensembles, where multiple onsets are close but not perfectly synchronous.

Novelty detection systems follow the standard architecture described in section 1.1, whereby the signal is reduced to a commonly 1-dimensional feature representation, the novelty function[1], followed by a simple detection or classification stage, where peaks of the novelty function are picked to determine the position of onsets.

In music with well defined and separated events, onsets might be fully characterized by amplitude increases in the time-domain signal. Therefore, simple functions such as envelope following or local energy act as rudimentary novelty functions. However, even for such signals, these methods can fail to characterize amplitude increases, instead emphasizing large absolute energy/envelope values. The characterization of increases can be achieved by means of the first-order derivative of the energy function with regards to time, which, as can be seen in the middle plot of Fig. 7.5 produces sharp peaks during bursts of energy. However, the figure also shows how the third event in the sequence virtually disappears from the novelty function as a result of its low energy. To address this issue, (Klapuri 1999) proposes considering that our ability to perceive these increases depends on the context in which they are heard. This work coarsely approximates the ear's loudness response, by making detectable changes in energy proportional to the overall energy of the sound, such that:

$$\frac{\partial E(m)/\partial m}{E(m)} = \frac{\partial log\,(E\,(m))}{\partial m} \qquad (7.8)$$

where E is the local energy function and m is an index of discrete time. The resulting function, seen in the bottom plot of Fig. 7.5, is much better at emphasizing low-energy onsets such a the third event in the sequence, but also susceptible to noisy behavior when energy approximates zero, as can be observed after the fifth and

---

[1] Also known as onset detection function, or onset strength signal.

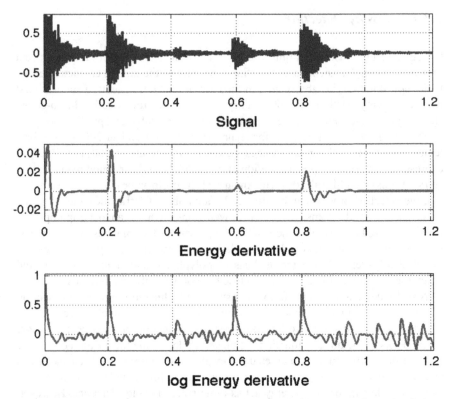

Fig. 7.5 Novelty detection using energy-based methods: audio signal (*top*), first-order derivative of the local energy (*middle*), and log of the energy derivative (*bottom*). The *horizontal axis* represents time in seconds

final event in the sequence. Thus, combining the derivative of the log-energy with an absolute energy threshold yields optimal results.

The above time-domain methods ignore the fact that new events are often well localized in frequency, and that considering changes of energy in the signal as a whole might obscure important changes occurring in specific frequency regions. Therefore, we can extend the idea of using first-order derivatives to characterize change separately in specific frequency bands. The resulting function is known as spectral flux (Bello et al. 2005):

$$SF_R(m) = \frac{2}{N} \sum_{k=0}^{N/2} H(|X_k(m)| - |X_k(m-1)|) \qquad (7.9)$$

where $H(x) = (x + |x|)/2$ is a half-wave rectification function that only takes energy increases into account. Some implementations either square or log the rectified flux before summation, in order to de-emphasize or emphasize, respectively, low energy events. Furthermore, the summation across bands, which is common in the

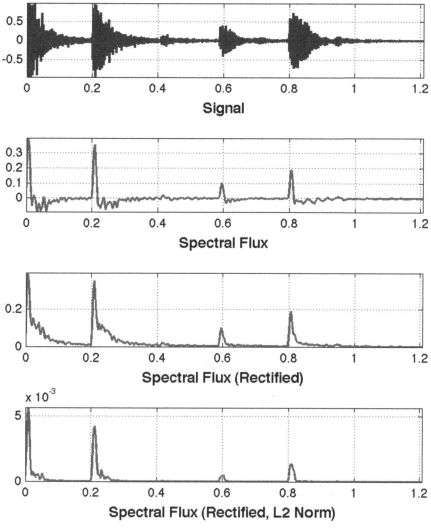

Fig. 7.6 Novelty detection using spectral flux methods: audio signal (*top*), spectral flux (*middle top*), rectified spectral flux (*middle bottom*), and normalized, rectified spectral flux (*bottom*). The *horizontal axis* represents time in seconds

context of novelty detection, might be undesirable in the context of transcription or the characterization of rhythm similarity, where being able to distinguish between events occurring at different frequencies is an asset. Figure 7.6 shows the spectral flux function with and without rectification, and with squaring prior to summation.

The spectral flux approach can be extended beyond magnitude information by taking into account deviations from predicted phase. Consider for example the ideal scenario where each channel k of the Fourier analysis encodes information about a single sinusoid in steady-state. In this scenario, we can expect the first-order phase difference to remain the same at consecutive time instants, resulting on the following prediction for the phase at time m:

$$\hat{\phi}_k(m) = princarg\left(2\phi_k(m-1) - \phi_k(m-2)\right) \tag{7.10}$$

where $princarg$ is a phase unwrapping function. If we predict the complex DFT as $\hat{X}_k(m) = |\hat{X}_k(m)|\, e^{j\hat{\phi}_k(m)}$, such that $|\hat{X}_k(m)| = |X_k(m-1)|$, then we can define:

$$CD(m) = \frac{2}{N}\sum_{k=0}^{N/2} RCD_k(m) \tag{7.11}$$

$$RCD_k(m) = \begin{cases} |X_k(m) - \hat{X}_k(m)| & \text{if } |X_k(m)| \geq |X_k(m-1)| \\ 0 & \text{otherwise} \end{cases} \tag{7.12}$$

The resulting complex-domain spectral flux computation is sensitive to pitch-based and soft onsets (Bello 2003).

Once a novelty function is computed, the positions of onsets in the signal can be obtained by identifying peaks in the function. To facilitate peak-picking, novelty functions are typically smoothed, to decrease jagedness; normalized, such that onset detection can be robustly done across signals; and thresholded, so as to eliminate spurious peaks. Thresholding is usually adaptive so as to minimize sensitivity to signal level changes (due, for example, to changes in the dynamics of a given performance), and can be done using local mean or median computations. Figure 7.7 shows that, after these processes have been applied to the novelty function (depicted in black), peak-picking reduces to selecting local maxima above the threshold (red line).

7.3.2   Tempo, Beats and Periodicity

Novelty functions, by characterizing the occurrence of new events in the signal, also encode information about the patterns of temporal organization that define the rhythm of a musical piece. Furthermore, those patterns are often repetitive at different time scales, so detecting periodicities on the novelty function can aid characterizing the tempo, beat and metrical structure of the piece, and thus help with music segmentation, classification and similarity tasks. Yet, periodicity detection is non trivial, as most of the repetitive patterns we want to identify are quasi-periodical, instead of perfectly oscillatory. Also, there are often multiple periodicities associated with a fundamental period, and multiple fundamental periods present in the signal. Finally, the presence of transient noise, temporal variations on the periodic structures we wish to analyze, and ambiguous and often misleading events, all add to the complexity of the task.

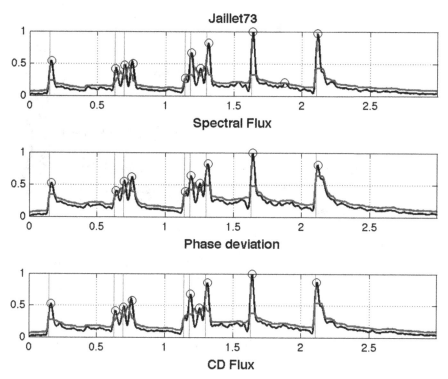

Fig. 7.7 Onset detection using multiple novelty functions: spectral flux (*top*), phase deviation (*middle*), and complex-domain spectral flux (*bottom*). The *horizontal axis* represents time in seconds. Selected onsets are marked with a *circle*. True onsets are marked with a *green vertical line*

There are numerous methods for analyzing the periodicity of signals, but here we will focus on using the discrete Fourier transform (DFT), a well-known and efficient strategy that has been very popular in rhythm analysis. The DFT of an N-long segment of the input signal, in this case a novelty function, can be seen as a linear combination of a predefined set of complex sinusoids with evenly-distributed frequency values. The spectral magnitude corresponds to the weights of the linear combination, while the spectral phase corresponds to individual offsets in time applied to each sinusoid.

Therefore, it follows that strong periodicities in the input signal result in high-valued peaks in the magnitude spectrogram. Furthermore, complex, real-world signals oscillate at multiple frequencies simultaneously, resulting on several spectral peaks mostly, but not exclusively, located at integer multiples of the main or fundamental periodicity. In other words, high spectral peaks will tend to be spaced in frequency intervals close to the fundamental frequency. Yet if a periodicity is strong enough, like the steady beat in the electronic music sample in Fig. 7.8, then a high peak will dominate consecutive magnitude spectra, as is clearly visible on the right-hand plot in the figure. Detecting this periodicity by peak-picking the spectrogram, results on an estimation of the local tempo of the piece (see (Gouyon et al. 2006) for a comprehensive review of tempo estimation methods).

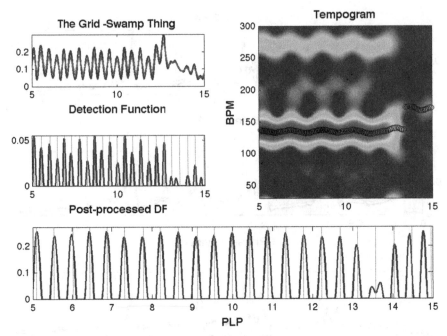

Fig. 7.8 Tempo estimation and tracking using the approach in (Grosche and Muller 2011): Electronic music example. The *horizontal axis* represents time in seconds

The magnitude spectra, however, only tells half the story. Beats not only occur at a certain frequency, the pulse or tempo, but at certain temporal locations in the signal. Yet timing can only be recovered from the phase of the DFT analysis, not its magnitude. In (Davies and Plumbley 2007), once the main periodicity has been identified, the authors convolve the corresponding periodic pattern (in their case a comb filter rather than a sinusoid) with the novelty function. The maximum of the convolution is indicative of the phase offset needed to recover the position of events. In (Grosche and Muller 2011), the authors construct a sinusoidal kernel using the frequency and phase of the maximum spectral peak, and then overlap-add with adjacent kernels to construct a new, beat or pulse-specific detection function, which they term the "predicted local pulse" (PLP) curve. This function can in turn be peak-picked for beat tracking. The bottom plot of Fig. 7.8 shows both the curve (blue line) and annotated beat positions (green) for the electronic music sample discussed above. Of course, the analysis is much harder for rhythmically-complex signals with less-steady beats, often resulting on octave and localization errors. Figure 7.9 shows such a case for an excerpt of Herbie Hancock's *Cantaloupe Island*.

7.3.3 Rhythm Similarity and Classification

In the above section, the focus was on detecting short-term rhythmic periodicities related to the beat structure. However, using longer windows in the DFT analysis

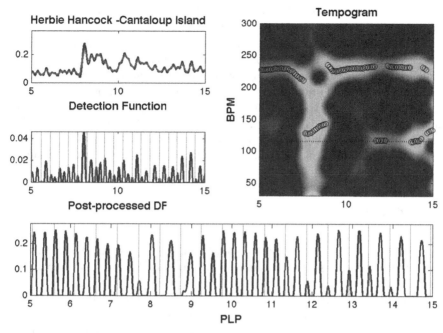

Fig. 7.9 Tempo estimation and tracking using the approach in (Grosche and Muller 2011): Jazz example. The *horizontal axis* represents time in seconds

can be used to characterize periodicities at the metrical level. It is important to note that there is a limit to how much window sizes can be increased without affecting the assumption of stationarity which is central to Fourier analysis and other similar techniques (Holzapfel et al. 2011).

The magnitude DFT of novelty functions, computed on windows somewhere between 4 and 10 s long, produces features that have been successfully applied to tasks such as genre classification and rhythmic similarity. This representation has several interesting properties that make it a good feature for describing rhythm. First, focusing on the magnitude means that the feature is invariant to temporal shifts of the rhythm pattern. In other words, as long as the information in the analysis window is mostly the same, where the pattern starts in the window has little to no influence on the magnitude spectrum. However, phase cancellations between the Fourier sinusoidal bases means that the process is still sensitive to sequential ordering (Peeters 2011). For example a 22211 *ChaChaCha* pattern has a different magnitude spectra than a 21221 *Rumba* pattern, despite both of those patterns having the same distribution of inter-event intervals. Peeters has clearly demonstrated (Peeters 2011) that this sensitivity, which is an asset when we are trying to differentiate between rhythmic patterns, extends to metrical structure, groove and swing.

Further, by computing the DFT using a logarithmic frequency scale, tempo changes become linear shifts in frequency, instead of the rescaling of spectral peak positions that occurs in standard DFT analysis. This combined with low frequency resolution results in partial tempo invariance, as large as the frequency range of DFT bins. In (Pohle et al. 2009), this log2 DFT analysis is applied to novelty functions across multiple frequency bands, then aggregated through time via a median operation, resulting in a rhythm feature known as *onset patterns*. The authors demonstrate robustness for the task of rhythm-based genre classification, and show how these features can contribute to improving timbre-based measures of music similarity. (Holzapfel et al. 2011) further improves on this representation by systematically exploring its parameter space, and by adding a processing layer, based on the scale transform (Holzapfel and Stylianou 2009), to introduce full tempo invariance in the representation. This additional in-variance is demonstrated to be useful for the analysis of Turkish and Cretan music showing significant, within-style, tempo changes.

7.4 Tonality

Western tonal music, to which most popular music belongs, is based on a system, *tonality*, that arranges sounds according to pitch relationships into inter-dependent spatial and temporal structures. Many music theorists consider tonality to be the main foundation of western music (see (Lerdahl 2001; Huron 2006) for relevant discussions). Moreover, empirical evidence suggests that untrained listeners intuitively expect and perceive basic patterns of tonal organization, almost as if they were born with this understanding (Krumhansl 1990; Janata et al. 2002).

The chord is the most fundamental unit of structure of the tonality system. Its identity and function are determined both by its constituent pitch classes and also by the syntactical context in which the chord occurs. Contextual significance is a consequence both of the chords position relative to other chords within a temporally organized sequence and also to its functional relationships to other chords and (ultimately) to an underlying tonic. Because the chord as a structure is comprised essentially of pitches, its contextual identity is typically strong enough to be perceived through changes in instrumentation, the addition or deletion of individual notes, and even the mixing in of non-pitched sounds or noises.

Machine listening research in tonal analysis has focused on the development of signal processing techniques for the detection of pitch class information in music, notably the so-called *chroma* features. Combinations of these signal representation with graphical models can be used for the automatic detection of chords and chord sequences. Further, a variety of techniques for feature post-processing and pairwise distance computation can be used to identify covers or remakes based on tonal content. The following provides more details about these techniques.

7.4.1 Chroma Features

The helical model of pitch perception (Shepard 1999), depicted in the left side of Fig. 7.10, exists in a space defined by the two dimensions of height and chroma. While height characterizes the perceived pitch increases concomitant with increases

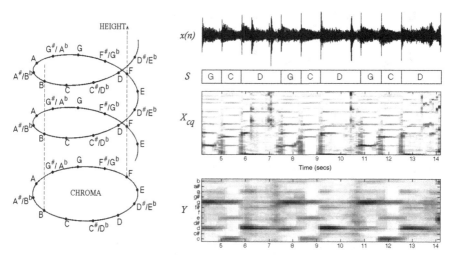

Fig. 7.10 The helical model of pitch perception (*left*), showing the two dimensions of height and chroma; and chromagram computation for an audio excerpt from Michael Chapman's *Another Crossroads* (*right*). The underlying chord sequence and signal's constant Q transform are shown in the *middle* plots

on the frequency of a sound, chroma relates to the cyclical perception of pitch as it moves from one octave to the other. According to this model, sounds whose frequencies are separated by an integral number of octaves occupy the same position in the chroma dimension.

There are a number of techniques for computing chroma features from audio signals, usually based on the warping and wrapping of the signal's magnitude spectrum. One such technique, illustrated in the right side of Fig. 7.10, uses the constant Q transform (Brown 1991), a spectral analysis where frequency-domain channels are logarithmically spaced thus closely resembling the frequency resolution of the human auditory system. The constant Q transform Xcq of a digital signal $x(n)$ can be calculated as:

$$X_{cq}(k) = \sum_{n=0}^{N(k)-1} w(n,k)x(n)e^{-j2\pi f_k n} \qquad (7.13)$$

where the analysis window w and its length N are functions of the frequency bin position k. The center frequency of the kth bin, $f_k = 2^{k/\beta}f_{min}$, is defined according to the frequencies of the equal-tempered scale. β is the number of bins per octave, thus defining the resolution of the analysis, and f_{min} defines the starting point of the analysis in frequency. From X_{cq}, the chroma vector for a given frame n can be calculated as:

$$y(b) = \sum_{i=0}^{O-1} |X_{cq}(b+i\beta)| \qquad (7.14)$$

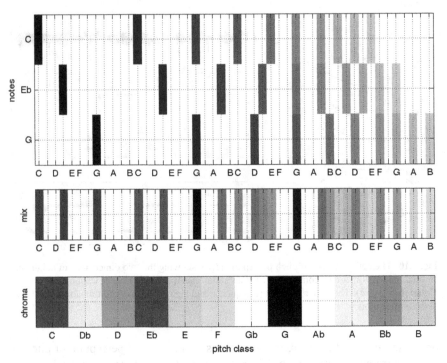

Fig. 7.11 Chroma computation for a synthesized C minor chord: spectra of individual notes (*top*), spectrum of the mixture (*middle*), and resulting chroma feature vector (*bottom*) showing the effects of overtone noise

where $b \in 1, \beta$ is the chroma bin number, and O is the total number of octaves considered. Just as with the short-time Fourier transform (STFT), chroma vectors can be calculated sequentially on partially overlapped blocks of signal data, resulting in a *chromagram*.

Chroma features, however, are not without limitations. Figure 7.11 shows the computation of a chroma vector corresponding to a C minor chord. The top plot shows the ideal frequency spectra of the notes composing the chord: C, Eb and G. Each spectrum contains energy only at the frequencies corresponding to the integer multiples of the fundamental frequency, such that the overtones' amplitude values decrease exponentially, visible in gray scale in the figure. The horizontal axis represents log-2 frequencies labeled using the pitch classes of the C major scale, to facilitate the analysis. Up to 10 harmonics are shown per plot.

It can be observed that the frequencies of the first, second, fourth and eight harmonics correspond to the nominal pitch class of the note. However, other harmonics correspond to different pitch classes at intervals such as the perfect fifth (3rd and 6th harmonic), major third (5th and 10th) and minor seventh (7th) from the fundamental frequency. Therefore, when the contributions of all harmonics are mixed (middle plot) and folded into one octave (bottom plot), the resulting representation contains energy in pitch classes other than the ones composing the chord. In the case of a minor chord, such as the one in the figure, this includes energy in pitch classes corresponding to both the major and minor third intervals, potentially causing confusion at later stages of the analysis. Beyond the synthesized example in the figure, real sounds have noisier and more complex magnitude spectra that results in even more ambiguous representations. This overtone "noise" is a reminder that, despite being a harmonic representation, chroma features are highly influenced by the timbre of the mixed sounds. These features, typically computed at fast frame rates, are also susceptible to the influence of short-term events such as transient noise, percussive sounds, even passing melody or bass notes.

There are potential solutions to these problems, including bandpass filtering and/ or whitening the log-2 spectrum, as well as low-pass filtering of chroma sequences over time. In the context of chord recognition, the work in (Cho et al. 2010) and (Cho and Bello 2014) has demonstrated that finding the optimal combination of chroma post-processing techniques has a dramatic impact on performance, significantly more so than the use of sophisticated machine learning strategies for analysis.

7.4.2 Chord Recognition

Despite the shortcomings described above, when a certain chord plays, the chroma features of the associated blocks of signal data show a concentration of energy in the pitch classes corresponding to the notes of the played chord. These emerging patterns and their sequential organization can be learned and recognized using graphical models, of which hidden Markov models (HMM) are a popular example. HMM are a class of dynamic models that provide a robust solution to the analysis of processes such as speech or music, which can be represented as a sequence of unknown, or *hidden*, states (e.g. the underlying chord progression) that can be characterized only from a set of observations (the chroma vectors y) (Rabiner 1989). These models are based on the Markov assumption, i.e. the probability of being at a given state only depends on the previous state of the process. A model with a finite set of states $S = \{s_j\}$, is defined by the set of parameters $\lambda = \{B, A, \pi\}$, where:

- $B = \{b_j\}$ is the state to observation probability distribution for each state j
- $A = \{a_{ij}\}$ is the state transition probability for every pair of states i, j
- $\pi = \{\pi_j\}$ is the initial state distribution

When the observations are continuous then each bj can be defined as a mixture of M multidimensional Gaussians:

$$b_j(y) = \sum_{m=1}^{M} w_{jm} \mathcal{N}[y; \mu_{jm}; \Sigma_{jm}] \qquad (7.15)$$

where w_{jm} is the mixing coefficient of the mth Gaussian with mean μjm and covariance matrix Σ_{jm}. Given the observation sequence $Y = y_1 \dots y_K$, the problem of hidden Markov models is to find the set of parameters λ and the state sequence $S = s_1 \dots s_K$ that maximize the joint probability:

$$p(Y,S) = \prod_{k=1}^{K} p(S_k \mid S_{k-1})p(y_k \mid S_k) \qquad (7.16)$$

The conditional probability $p(s_k \mid s_{k-1})$ contains the Markov assumption, while $p(y_k \mid s_k)$ implies independence between the observation vectors y_k. There are well-known solutions to this optimization problem (Rabiner 1989), with the optimal sequence S becoming the representation of the audio signal from which Y is derived.

The combination of chroma features and graphical models for chord sequence estimation was pioneered in (Sheh and Ellis 2003), but this early work attempted optimizing too large a parameter space with too little chord-annotated data, resulting in poor performance. Since then, the community has produced significant amounts of annotated data (Harte et al. 2005; Burgoyne et al. 2011; Cho and Bello 2011), and has experimented with many variations of this formulation including hand designed parameters based on basic musicianship (Bello et al. 2005; Papadopoulos and Peeters 2007), models trained with MIDI data (Lee 2007), and replacing standard HMM with hierarchical models (Khadkevich and Omologo 2009), conditional random fields (Burgoyne et al. 2007), Bayesian networks (Mauch and Dixon 2010b), structured support vector machines (Weller et al. 2009) and convolutional neural networks (Humphrey et al. 2013), to cite only a few examples. However, the level of sophistication in machine learning methods and training strategies which is in display in the literature has rarely being matched in feature design, a notable observation given that, as was mentioned at the end of the previous section, it has been experimentally shown that feature design and post-processing have a larger impact on performance (Cho et al. 2010; Cho and Bello 2014).

At the time of writing, state of the art systems were able to identify major and minor triads, as well as the absence of chords, with accuracies of up to 80% on datasets of hundreds of Western pop music tracks. Accuracies decrease to less than 70% when identifying more complex, 4-note chords and inversions as well (Mauch and Dixon 2010a; Ni et al. 2012; Cho and Bello 2014). While far from perfect, these approaches are the best examples of automatic music transcription systems currently in existence.

7.4.3   Tonal Similarity and Cover Song Identification

As was the case for timbral and rhythmic content, we can propose measures of "harmonic similarity" between audio tracks, and search music collections accordingly. Similarity in this case, is a function of tracks having common harmonic subsequences, regardless of changes of instrumentation, dynamics, tempo, and key. Performance is typically validated in the context of cover-song identification, a query-by-example task where, given a recorded version of a musical work, the goal is to find all other versions or covers of the same work. The assumption made by these approaches is usually that musical works are typified by specific melodic or harmonic sequences.

There are numerous approaches in the literature for automatic cover song identification. The vast majority of methods use some form of normalized chroma
features to focus on pitch content and introduce partial invariance to timbre and
dynamics. Yet, there are many competing approaches to achieving tempo and
transposition invariance. For example, one group of techniques utilizes symbolic
chord sequences, rotated using transposition estimation heuristics to a common
key, and compared using sequence alignment tools popular in bioinformatics to
achieve invariance to tempo and temporal location (Lee 2006; Bello 2007; Martin
et al. 2012). However, it has been found that error propagation resulting from chord
estimation has a negative effect on performance, and that better results can be found
when operating directly on the signal's chromagram (Serra et al. 2008).

The groundbreaking work in (Serra et al. 2008; Serrà et al. 2009) proposes
multiple approaches, including pairwise, binary measures of chroma similarity that
take into account the optimal transposition between sequences, as well as methods
based on dynamic time warping or cross-recurrence quantification, to find common
subsequences despite variations in tempo. Yet, for most of this work, invariances
are introduced at the expense of costly pairwise similarity computations that have
to be run during search time and thus preclude these approaches from scaling well
to collection of hundreds of thousands, even millions of tracks.

As an astute solution to this problem, (Bertin-Mahieux and Ellis 2012) proposes
a method based on computing the 2-dimensional Fourier Magnitude Coefficients
(2D-FMC) of long subsequences of beat-synchronous chroma features. Shifts in
time—e.g. due to comparing similar sequences starting at different metrical positions and pitch class—e.g. due to different keys—are encoded in the phase component of the DFT. Thus invariance is achieved by preserving only the magnitude of
the representation. 2D-FMC are computed on a moving window of 75 beat chroma
vectors. Then, the median is taken across all vectors and PCA is used for dimensionality reduction, resulting on a single vector for the entire track. By encoding
invariances in the feature vector, the approach can identify cover songs by means
of a simple nearest-neighbors search using Euclidean distances. This is an efficient
process that scales well to the analysis of millions of tracks, as demonstrated by
tests on the Million Song Dataset (Bertin-Mahieux et al. 2011), the largest music
audio collection currently available for research. In (Humphrey et al. 2013), this approach is expanded by incorporating a convolutional sparse coding stage for feature
learning, which increases the separability of tracks in the representation space, and
supervised dimensionality reduction, using linear discriminant analysis, to recover
an embedding where covers are close together in an Euclidean sense. These additions significantly improve the retrieval of covers at the top of the ranked list.

7.5 Conclusions

This chapter has briefly reviewed a number of standard methods for the machine
listening of music, with a strong focus on signal feature extraction. More specifically, it describes common theories and techniques behind the automatic analysis

of timbral, rhythmic and tonal content in music signals, as well as their classification into predefined categories, and their organization according to some notion of similarity. These methods constitute the backbone of current research in music information retrieval and related fields, and are in wide distribution via toolboxes and large collections of pre-computed feature sets such as the Million Song Dataset.

Beyond introducing the techniques, this chapter has also attempted to shed some light on the thinking behind these methods, their embedded assumptions and limitations. It cannot be stressed enough that making the right implementation choices and finding the optimal parameterization matters greatly. Many of these decisions are task or music specific, and will thus have to be considered on a case by case basis, but they are never without consequences, and the researcher or developer that ignores them does so at their peril. In the past it has been possible to throw ever more powerful machine learning techniques at the same set of features and improve results. But recent trends, as exemplified by results on the Music Information Retrieval Evaluation eXchange (MIREX), show that progress on many standard tasks is slowing down if not altogether stalling, and that approaches can benefit from new and improved signal representations. This is particularly important for tasks such as mood estimation, autotagging or automatic remixing, where it is unclear that existing feature representations are enough. Yet, designing new features and the best way to implement them is far from trivial. We have seen already how even for well-known problems like chord estimation, and well-known features like chroma, the community has taken the best part of a decade to come up with an indepth understanding of how to explore the range of design choices.

One interesting avenue for development is to turn to feature learning as an alternative to feature design, thus leveraging data in lieu of, or as a complement to, domain knowledge. In fact, it could be argued that current feature extraction methods are specific instances of a general architecture stacking layers of subsampling operations and affine transforms with parameters set by hand. Instead, those parameters can, and have been, learned from data in the context of research on non-negative matrix factorization (Smaragdis and Brown 2003; Weiss and Bello 2011) or sparse coding (Daudet 2006; Henaff et al. 2011), to name only the most visible examples. However it is only recently, and following relevant developments in machine learning and other fields of computational perception, that these specific feature learning layers are starting to be seen as elements of deep architectures integrating aspects of feature extraction and classification (Bengio 2009), whose parameters are tuned using a combination of unsupervised and supervised learning approaches. Work on deep learning applied to the machine listening of music is still in its infancy, but is already showing promise for applications in music classification (Lee et al. 2009; Hamel and Eck 2010; Nam et al. 2012), instrument identification (Humphrey et al. 2011), music similarity (Schluter and Osendorfer 2011), mood estimation (Schmidt and Kim 2011), chord recognition (Humphrey et al. 2012), and drum pattern analysis (Battenberg and Wessel 2012). These developments hold transformative potential for machine listening in general, and the computational analysis of music in particular, and will no doubt have an important role to play in the future of the field.

References

Agawu K (2012) Trends in African musicology: a review article. EthnoMusicol 56(1):133–140

Aucouturier JJ (2006) Ten experiments on the modelling of polyphonic timbre. PhD thesis, University of Paris 6, France

Aucouturier, J.-J., Defreville, B. and Pachet, F. *The bag-of-frame approach to audio pattern recognition: A sufficient model for urban soundscapes but not for polyphonic music*. Journal of the Acoustical Society of America, 122(2):881–91, 2007.

Bamberger JS, Hernandez A (2000) Developing musical intuitions: a project-based introduction to making and understanding music. Oxford University Press, New York

Barbedo JGA (2012) Instrument recognition. In: Li T, Ogihara M, Tzanetakis G (eds) Music data mining. CRC Press, Boca Raton, Florida, USA

Battenberg E, Wessel D (2012) Analyzing drum patterns using conditional deep belief networks. In: ISMIR, pp 37–42

Bello JP (2003) Towards the automated analysis of simple polyphonic music: a knowledge-based approach. PhD thesis, Department of Electronic Engineering, Queen Mary University of London

Bello JP (September 2007) Audio-based cover song retrieval using approximate chord sequences: testing shifts, gaps, swaps and beats. In: Proceedings of the 8th international conference on music information retrieval (ISMIR-07). Vienna, Austria, September 2007.

Bello JP, Daudet L, Abdallah S, Duxbury C, Davies M, Sandler MB (September 2005) A tutorial on onset detection in music signals. IEEE Trans Speech Audio Process 13(5):1035–1047 (Part 2)

Bengio Y (January 2009) Learning deep architectures for AI. Found Trends Mach Learn 2(1):1–127

Berenzweig A (2007) Anchors and hubs in audio-based music similarity. PhD thesis, Columbia University, New York

Berenzweig A, Logan B, Ellis D, Whitman B (2003) A large-scale evaluation of acoustic and subjective music similarity measures. In: Proceedings of the international conference on music information retrieval, Baltimore

Bertin-Mahieux T, Ellis DPW (2012) Large-scale cover song recognition using the 2D Fourier transform magnitude. In: The 13th international society for music information retrieval conference, pp 241–246

Bertin-Mahieux T, Ellis DPW, Whitman B, Lamere P (2011) The million song dataset. In: Proceedings of the 12th international conference on music information retrieval (ISMIR 2011)

BMAT (2013) http://www.bmat.com/ Accessed July 20, 2013

Brown J (1991) Calculation of a constant Q spectral transform. J Acoust Soc Am 89(1):425–434

Burgoyne JA, Pugin L, Kereliuk C, Fujinaga I (2007) A cross-validated study of modelling strategies for automatic chord recognition in audio. In: ISMIR, pp 251–254

Burgoyne JA, Wild J, Fujinaga I (2011) An expert ground truth set for audio chord recognition and music analysis. In: Proceedings of the conference of the international society for music information retrieval (ISMIR), Miami, FL, pp 633–638

Cho T, Bello JP (2011) A feature smoothing method for chord recognition using recurrence plots. In: Proceedings of the conference of the international society for music information retrieval (ISMIR)

Taemin Cho; Bello, J.P., "On the Relative Importance of Individual Components of Chord Recognition Systems," *Audio, Speech, and Language Processing, IEEE/ACM Transactions on*, vol.22, no.2, pp.477,492, Feb. 2014

Cho T, Weiss RJ, Bello JP (July 2010) Exploring common variations in state of the art chord recognition systems. In: Proceedings of the sound and music computing conference (SMC-10), Barcelona

Cook PR (2001) Music, cognition, and computerized sound: an introduction to psychoacoustics. The MIT Press, Cambridge, MA, USA.

Daudet L (September 2006) Sparse and structured decompositions of signals with the molecular matching pursuit. IEEE Trans Audio Speech Lang Process 14(5):1808–1816

Davies MEP, Plumbley MD (2007) Context-dependent beat tracking of musical audio. IEEE Trans Audio Speech Lang Process 15(3):1009–1020

Gouyon F, Klapuri A, Dixon S, Alonso M, Tzanetakis G, Uhle C, Cano P (2006) An experimental comparison of audio tempo induction algorithms. IEEE Trans Audio Speech Lang Process 14(5):1832–1844

Gracenote (2013) http://www.gracenote.com/music/

Grey JM (1975) An exploration of musical timbre. PhD thesis, Department of Music, Stanford University

Grosche P, Muller M (2011, to appear) Extracting predominant local pulse information from music recordings. IEEE Trans Audio Speech Lang Process

Hamel P, Eck D (2010) Learning features from music audio with deep belief networks. In: ISMIR, Utrecht, pp 339–344

Harte C, Sandler MB, Abdallah SA, Gómez E (2005) Symbolic representation of musical chords: a proposed syntax for text annotations. In: Proceedings of the conference of the international society for music information retrieval (ISMIR), London, pp 66–71

Henaff M, Jarrett K, Kavukcuoglu K, LeCun Y (2011) Unsupervised learning of sparse features for scalable audio classification. In: Proceedings of international symposium on music information retrieval (ISMIR'11)

Herrera P, Klapuri A, Davy M (2006) Automatic classification of pitched musical instrument sounds. In: Klapuri A, Davy M (eds) Signal processing methods for music transcription. Springer, New York, pp 163–200

Hockman J, Bello JP, Davies MEP, Plumbley M (September 2008) Automated rhythmic transformation of musical audio. In: Proceedings of the International Conference on Digital Audio Effects (DAFX-08), Espoo

Holzapfel A, Stylianou Y (2009) A scale transform based method for rhythmic similarity of music. In: Proceedings of the IEEE International Conference on Acoustics, Speech, and Signal Processing (ICASSP), Taipei

Holzapfel A, Flexer A, Widmer G (2011) Improving tempo-sensitive and tempo-robust descriptors for rhythmic similarity. In: Proceedings of SMC 2011, Conference on Sound and Music Computing

Honing H (2012) The structure and interpretation of rhythm in music. In: Deutsch D (ed) The psychology of music, 3rd edn. Academic Press, London, pp 369–404

Humphrey E, Glennon A, Bello JP (December 2011) Non-linear semantic embedding for organizing large instrument sample libraries. In: Proceedings of the IEEE international conference on machine learning and applications (ICMLA-11), Honolulu

Humphrey E, Cho T, Bello JP (2012) Learning a robust tonnetz-space transform for automatic chord recognition. In: Proceedings of the IEEE international conference on acoustics, speech, and signal processing (ICASSP-12). Kyoto, Japan. May, 2012

Humphrey E, Bello JP, LeCun Y (December 2013) Feature learning and deep architectures: new directions for music informatics. J Intell Inf Syst 41(3):461–481

Huron D (2006) Sweet anticipation: music and the psychology of expectation. MIT Press Cambridge, MA, USA.

Janata P, Birk JL, Van Horn JD, Leman M, Tillmann B, Bharucha JJ (2002) The cortical topography of tonal structures underlying western music. Science 298:2167–2170

Janata P, Tomic ST, Haberman JM (2012) Sensorimotor coupling in music and the psychology of the groove. J Exp Psychol Gen 141(1):54

Jehan T (2005) Creating music by listening. PhD thesis, Massachusetts Institute of Technology, MA, USA, September

Khadkevich M, Omologo M (2009) Use of hidden markov models and factored language models for automatic chord recognition. In: Proceedings of the conference of the International Society for Music Information Retrieval (ISMIR), Kobe, Japan, pp 561–566

Klapuri A (1999) Sound onset detection by applying psychoacoustic knowledge. In: Proceedings of the IEEE International Conference on Acoustics, Speech, and Signal Processing (ICASSP), Washington, D.C., USA, pp 3089–3092

Kolinski M (1973) A cross-cultural approach to metro-rhythmic patterns. Ethnomusicology 17(3):494–506

Krumhansl CL (1990) Cognitive foundations of musical pitch. Oxford University Press, New York

Lee K (2006) Identifying cover songs from audio using harmonic representation. In: MIREX task on audio cover song ID

Lee K (May 2007) A system for chord transcription, key extraction, and cadence recognition from audio using hidden Markov models. PhD thesis. Stanford University, CA, USA, May 2007

Lee H, Largman Y, Pham P, Ng AY (2009) Unsupervised feature learning for audio classification using convolutional deep belief networks. In: Advances in Neural Information Processing Systems (NIPS), pp 1096–1104

Lerdahl F (2001) Tonal pitch space. Oxford University Press, New York

Lewis AC (2007) Rhythm: what it is and how to improve your sense of it. RhythmSource Press, San Francisco

London J (2012) Hearing in time. Oxford University Press, New York

Martin B, Brown DG, Hanna P, Ferraro P (2012) Blast for audio sequences alignment: a fast scalable cover identification tool. In: ISMIR, pp 529–534

Mauch M, Dixon S (2010a) Approximate note transcription for the improved identification of difficult chords. In: ISMIR, pp 135–140

Mauch M, Dixon S (2010b) Simultaneous estimation of chords and musical context from audio. IEEE Trans Audio Speech Lang Process 18(6):1280–1289

McFee B, Barrington L, Lanckriet G (2012) Learning content similarity for music recommendation. IEEE Trans Audio Speech Lang Process 20(8):2207–2218

Nam J, Herrera J, Slaney M, Smith JO (2012) Learning sparse feature representations for music annotation and retrieval. In: ISMIR, pp 565–570

Ni Y, McVicar M, Santos-Rodriguez R, Bie TD (2012) An end-to-end machine learning system for harmonic analysis of music. IEEE Trans Audio Speech Lang Process 20(6):1771–1783

Oppenheim AV, Schafer RW (2004) From frequency to quefrency: a history of the cepstrum. Signal Processing Mag IEEE 21(5):95–106

Papadopoulos H, Peeters G (2007) Large-scale study of chord estimation algorithms based on chroma representation and hmm. In: Content-Based Multimedia Indexing. 2007. CBMI'07. International Workshop on (IEEE), pp 53–60

Peeters G (2011) Spectral and temporal periodicity representations of rhythm for the automatic classification of music audio signal. Audio Speech Lang Process IEEE Trans 19(5):1242–1252

Pohle T, Schnitzer D, Schedl M, Knees P, Widmer G (2009) On rhythm and general music similarity. In: Proceedings of the Conference of the International Society for Music Information Retrieval (ISMIR), Kobe, Japan, pp 525–530

Rabiner LR (1989) A tutorial on HMM and selected applications in speech recognition. Proc IEEE 77(2):257–286

Ravelli E, Bello JP, Sandler M (April 2007) Automatic rhythm modification of drum loops. IEEE Signal Proc Lett 14(4):228–231

Schluter J, Osendorfer C (2011) Music similarity estimation with the mean-covariance restricted boltzmann machine. In: Machine Learning and Applications and Workshops (ICMLA), 2011 10th International Conference on (IEEE), vol 2, pp 118–123

Schmidt EM, Kim YE (2011) Learning emotion-based acoustic features with deep belief networks. In: Applications of Signal Processing to Audio and Acoustics (WASPAA), 2011 IEEE Workshop on (IEEE), pp 65–68

Schnitzer D, Flexer A, Schedl M, Widmer G (2012) Local and global scaling reduce hubs in space. J Mach Learn Res 13:2871–2902

Serra J, Gomez E, Herrera P, Serra X (2008) Chroma binary similarity and local alignment applied to cover song identification. IEEE Transactions on Audio, Speech and Language Processing. 16, 2008

Serrà J, Serra X, Andrzejak RG (September 2009) Cross recurrence quantification for cover song identification. New J Phys 11:093017, September 2009

Sheh A, Ellis D (October 2003) Chord segmentation and recognition using EM- trained hidden Markov models. In: Proceedings of the 4th International Conference on Music Information Retrieval (ISMIR-03). Baltimore, USA, pp 185–191

Shepard R (1999) Pitch perception and measurement. In: Cook P (ed) Music, cognition, and computerized sound. MIT Press, Cambridge, pp 149–165

Smaragdis P, Brown JC (2003) Non-negative matrix factorization for polyphonic music transcription. In: IEEE Workshop on Applications of Signal Processing to Audio and Acoustics, pp 177–180

Smith JO (2007) Mathematics of the discrete fourier transform (DFT): with music and audio applications. W3K http://books.w3k.org/

The Echonest (2013) http://the.echonest.com/ Accessed July 20, 2013

Toussaint G (2013) The geometry of musical rhythm: what makes a good rhythm good? CRC Press, Boca Raton, FL, USA.

Turnbull D, Barrington L, Torres D, Lanckriet G (2008) Semantic annotation and retrieval of music and sound effects. IEEE Trans Audio Speech Lang Proces 16(2):467–476

Tzanetakis G, Cook P (July 2002) Musical genre classification of audio signals. IEEE Trans Speech Audio Proces 10(5):293–302

Weiss RJ, Bello JP (2011) Unsupervised discovery of temporal structure in music. IEEE J Sel Top Signal Proces 5(6):1240–1251

Weller A, Ellis D, Jebara T (2009) Structured prediction models for chord transcription of music audio. In: Machine Learning and Applications, 2009. ICMLA'09. International Conference on (IEEE), pp 590–595

Wessel DL (1979) Timbre space as a musical control structure. Comp Music J 3(2):45–52

Widmer G, Dixon S, Goebl W, Pampalk E, Tobudic A (2003) In search of the Horowitz factor. AI Mag 24(3):111–130

Chapter 8
Making Things Growl, Purr and Sing

Stephen Barrass and Tim Barrass

" All our knowledge has its origin in the senses."

—Leonardo da Vinci

8.1 Introduction

The seminal book, Making Things Talk, provides instructions for projects that connect physical objects to the internet, such as a pet's bed that sends text messages to a Twitter tag (Igoe 2007). Smart Things, such as LG's recently released Wifi Washing Machine, can be remotely monitored and controlled with a browser on a Smart Phone. The Air-Quality Egg http://airqualityegg.com is connected to the Internet of Things where air-quality data can be shared and aggregated across neighborhoods, countries or the world. However, as Smart Things become more pervasive there is a problem that the interface is separate from the thing itself. In his book on the Design of Everyday Things, Donald Norman introduces the concept of the Gulf of Evaluation that describes how well an artifact supports the discovery and interpretation of its internal state (Norman 1988). A Smart Kettle that tweets the temperature of the water as it boils has a wider Gulf of Evaluation than an ordinary kettle that sings as it boils, because of the extra levels of indirection required to access and read the data on a mobile device compared to hearing the sound of the whistle.

In this chapter we research and develop interactive sonic interfaces designed to close the Gulf of Evaluation in interfaces to Smart Things. The first section describes the design of sounds to provide an emotional connection with an interactive couch designed in response to the Experimenta House of Tomorrow exhibition (Hughes et al. 2003) that asked artists to consider whether "the key to better living will be delivered through new technologies in the home". The design of the

---

S. Barrass (✉)
Faculty of Arts and Design, University of Canberra, Canberra, Australia
e-mail: stephen.barrass@canberra.edu.au

T. Barrass
Independent Artist and Inventor, Melbourne, Australia
e-mail: barrasstim@gmail.com

N. Lee (ed.), *Digital Da Vinci*, DOI 10.1007/978-1-4939-0536-2_8,
© Springer Science+Business Media New York 2014

couch to provide "both physical and emotional support" explores the principle of the hedonomics movement that proposes that interfaces to products should provide pleasurable experiences (Hancock et al. 2005). In the course of this project we developed a method for designing sounds that link interaction with emotional responses, modeled on the non-verbal communication between people and their pets (Barrass 2013). The popularity of ZiZi the Affectionate Couch in many exhibitions has demonstrated that affective sounds can make it pleasurable to interact with an inanimate object.

The next section explores the aesthetics of sounds in interfaces to Smart Things designed for outdoors sports and recreational activities. The first experiment investigates preferences between six different sonifications of the data from an accelerometer on a Smart Phone. The results indicate that recreational users prefer more musical sonic feedback while more competitive athletes prefer more synthetic functional sounds (Barrass et al. 2010). The development of interactive sonifications as interfaces to Smart Things was explored further in prototypes that synthesized sounds on Arduino microprocessors with various kinds of sensors and wireless communications. These prototypes also explored different materials that float and can be made waterproof, shapes that cue different manual interactions, and sonifications that convey different kinds of information. Technical issues with parallel data acquisition and sound synthesis on the Arduino microprocessor were addressed by extending the open source software to allow continuous real-time sound synthesis without blocking the simultaneous data acquisition from sensors (Barrass and Barrass 2013).

The final section describes the Mozzi software that grew out of these experiments. Mozzi can be used to generate algorithmic music for an artistic installation, wearable sounds for a performance, or interactive sonifications of solar panel sensors, on a small, modular and cheap Arduino, without the need for additional circuitry, message passing or external synthesizer hardware (Barrass 2013). The software architecture and Application Programming Interface (API) of Mozzi are described, with examples from an online tutorial that can guide further learning. The final section provides some examples of what can be done with Mozzi that include artistic installations and scientific applications that have been created by the Mozzi community.

8.2 Affective Sounds

The call for the Experimenta House of Tomorrow exhibition asked artists to question the effects of technology on domestic life in the future (Hughes et al. 2003). This led us to think about how the beeps of the microwave, dishwasher, and robot vacuum cleaner could be designed to be more pleasurable, engaging and communicative? What if things around the house that normally sit quietly made sounds? What kinds of sounds would furniture make, and why? To find out, we made a couch that whines for attention, yips with excitement, growls with happiness, and purrs with contentment. Would these interactive sounds increase the pleasure of

Fig. 8.1 ZiZi the Affectionate Couch at the Experimenta House of Tomorrow exhibition. Photo courtesy of Experimenta Media Arts

sitting on it, or increase the irritation we feel when bombarded with meaningless distractions? ZiZi the Affectionate Couch is an ottoman covered in striped fake fur, shown in Fig. 8.1.

A motion detector inside the couch senses movement up to 3 m away. Bumps with fur tufts on the surface suggest patting motions. The sensor signal varies in strength and pattern with the distance and kind of motion. The signal is analyzed by a PICAXE microprocessor to identify four states of interaction—no-one nearby, sitting, patting and stroking. When there is no-one nearby the couch is bored, and whines occasionally to attract attention. When someone sits on it, the couch expresses excitement with short yips, modeled on the response of a dog when it greets you. When the couch is patted, it expresses happiness through longer growls. When the couch is stroked at length, it expresses contentment by purring like a tiger. The design of sounds to convey these emotions was based on studies in which participants rated a 111 sounds on an Affect Grid, that has axes of Valence and Arousal (Bradley and Lang 2007). Valence is a rating from *displeasure* to *pleasure*, whilst Arousal is rated from *sleepy* to *excited*. The four states of interaction are shown plotted on the Affect Grid according to the emotion that should be conveyed, as shown in Fig. 8.2. The transition between states is shown by a path with a bead on it showing the current state. The sound palette is also plotted on the Affect grid, and samples that fall within the shading convey the affective response to that interaction. This Interactive Affect Design Diagram provides a new way to design sounds to convey emotion in response to interaction states. In future work the Interactive Affect Design Diagram will be tested as a way to design more complex sonic characters where positive valency sounds encourage some kinds of interaction, and negative valency sounds discourage others.

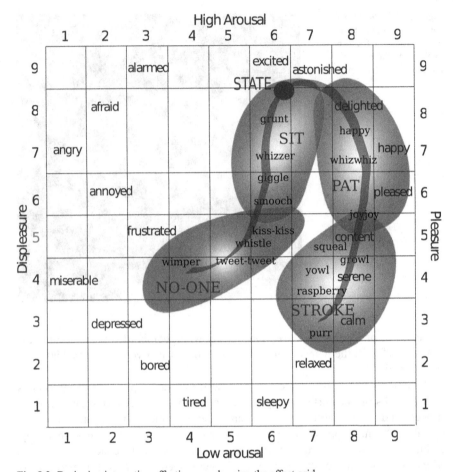

Fig. 8.2 Designing interactive affective sounds using the affect grid

The microprocessor is connected to a RAM chip that stores 16 mono sound samples. A four channel mixer and amplifier allows up to four of the samples to be combined to create more complex sounds and reduce repetition. The sounds are played through nine sub-woofer rumble-packs to produce purring vibrations and effects. The sounds can be heard at <http://stephenbarrass.com/2010/03/02/sounds-of-zizi/>.

8.3   Aesthetics of Sonic Interfaces

Mobile Apps that track health and fitness in outdoor activities are popular on Smart Phones. However the touchscreen interface is distracting and demanding during eyes-busy, hands-busy, outdoor activities. Although speech interfaces can provide

Fig. 8.3 Sweatsonics App on a smart phone

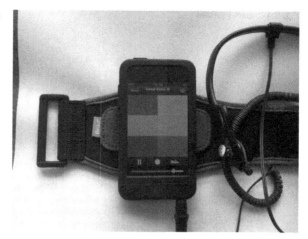

alarms and summaries, they cannot convey continuous multi-sensor data in real time. The sensors, data processing and communications built into Smart Devices led us to develop an App to explore the idea that data sonifications could provide continuous feedback for sports and recreational activities. The Sweatsonics App, shown in Fig. 8.3, was designed to study aesthetic preferences between six different sonifications of the 3 axis accelerometer on a Smart Phone. The phone was strapped to the arm of the participant, who listened through headphones whilst involved in a fitness activity of their own choosing, such as walking, jogging or tai-chi. During the activity, the time and duration of selections of different sonifications was logged.

The results showed preferences for the sine wave sonification (which sounds like a synthesizer or theremin), and a sonification with 3 musical instruments. The results from this pilot study raised many more questions about the aesthetics of sonification in sports and fitness. Were more competitive users choosing the sine wave which sounds like a medical device, whilst the more recreational users chose the more musical sonification? Was there a relationship with the energy of the activity, e. g. jogging compared to tai-chi? Were there age or gender differences? What is clear is that there are many factors that affect preferences between sonifications of the same sensor data with different users in different activities and contexts.

8.4   The Sonification of Things

During the experiments with the Sweatsonics App we found that users often knocked or dropped the Smart Phone, which is expensive to fix or replace, and that watery environments like rowing on the lake were particularly risky. Rugged casings can increase the robustness, but it can be difficult to retrieve a Smart Thing that has been dropped in the lake. The PICAXE microcontroller system with the additional sample

Fig. 8.4 Smart Flotsam

player that we made for the couch is larger than a Smart Phone, and too unwieldy to embed in a hand-sized object. However, the open source Arduino microprocessor has a capability to synthesize audio tones with variable frequency and duration. Arduino clones, with diminishing sizes, are designed to be embedded in smaller objects. This led to the choice of the Arduino in the next series of experiments with embedded sonifications as interfaces to Smart Things. The three prototypes, called Flotsam, Jetsam and Lagan, have a nautical theme motivated by our observations of the risk of dropping a Smart Phone in the lake in earlier experiments. Each prototype combines and explores a mixture of different materials, shapes, technologies, sounds, algorithms, and metaphors with a focus on robustness, waterproofing and floatability in a wet outdoor environment.

This first design was inspired by discussions about rowing, where the turning points in graphs of the acceleration of the skiff are used to understand and improve the smoothness and power of the stroke technique. The first prototype, called Flotsam, synthesises a "clicking" sound triggered by turning points in acceleration. The rate of clicks increases with the jerkiness of the movement, and stops when the movement is continuous, or stationary. This specialized Smart Thing is made of wood that floats and protects it from knocks. The two hollowed out halves are screwed together through a rubber gasket to make it watertight, as shown in Fig. 8.4.

Inside is an Arduino clone, called the Boarduino, which has a smaller 75 x 20 x 10 mm footprint. There are no knobs, buttons or screens. The synthesized sounds are transmitted to the outside with a Sparkfun NS73M FM radio chip tuned between 87–108 MHz. A 30 cm antennae wire wrapped around the inside of the casing broadcasts the signal 3–6 m. Flotsam could allow a team of rowers to listen to the turning points in the motion of their skiff through radio headphones, potentially

Fig. 8.5 Smart Jetsam

providing a common signal for synchronizing the stroke. Although the sounds come from the headphones, rather than the object itself, the direct relationship between the clicking and movement creates a sense of causality.

The click is synthesized with the built-in *tone*() command, set at 1 kHz frequency for 10 ms duration. In designing sounds on the Arduino it is important to understand that the *tone*() command blocks all other data acquisition and control functions until it completes. In testing we found that a click of 10 ms duration is audible without causing a noticeable interference with the other routines. In trials the FM transmission frequency drifted over time, which required the receiver to be re-tuned after 10–20 min, and sometimes it was difficult to find the signal again. However, the biggest problem was interference from commercial radio channels with much stronger signals scattered across the FM spectrum. Unscrewing the halves of the shell to re-tune the transmitter is inconvenient, but an external knob could compromise the robustness and waterproofing. The 9 V alkaline battery lasts 10 h, but unscrewing the halves to replace it is also inconvenient.

Flotsam demonstrated that useful sonifications can be designed with the tone() command on an Arduino that is small and cheap enough to be embedded in things used in demanding outdoor activities. It raised issues with the range of sounds that can be produced, and the technical problems of transmitting sounds from inside the object to the outside.

This next design iteration was motivated by the observation that it can be difficult to tell someone how to hold a piece of sporting equipment, such as an oar, or a tennis racquet. Could the Smart Thing itself guide the orientation and position of the user's grip? Jetsam has a twisted shape that can be held in different ways and orientations. This time we investigated haptic vibration as a mode of information feedback that allows the object to be sealed, waterproof and floatable.

Jetsam was carved from pumice which is a natural material formed from lava foam aerated with gas bubbles that can be found floating in the sea or lakes near volcanoes, as shown in Fig. 8.5. The exterior was sealed with several layers of polyester to make a smooth and toughened surface. The halves were hollowed out to make space for an Ardweeny clone that has a 40 x 14 x 10 mm footprint that is half the length of the Boarduino. The 3 axis accelerometer was soldered to the analog inputs on the Ardweeny, which was then inserted in the bottom half, while the 9 V battery was inserted into the top half. The two halves are connected by a screw-top (made from a plastic milk bottle) that seals but can be quickly and easily opened to replace the battery and re-program parameters on the microprocessor.

Mobile phone vibrators are glued inside at the top, on one side in the middle, and at the bottom, so that the haptic vibrations are localizable. The vibration is subsonic in the range from 0 to 20 Hz, after which the feeling becomes continuous. The first experiments with the tone() command to generate vibration effects used frequencies from 0–10 Hz but these could not be easily felt. This led to the development of a "pulse train" algorithm where each pulse was generated by a 1 kHz tone() with a duration of 50 ms which enabled the perception of variation in pulse rate between 0 and 10 Hz. The top vibrator pulses faster in linear steps as the x axis orientation goes from 0 to $+-90°$. The bottom end pulses in response to y axis orientation, and the middle pulses with the z axis. If the object is held vertical then the top pulses at the fastest rate. A slight horizontal tilt causes the bottom to pulse slowly as well. A tilt in all three directions produces pulses at different rates at the top, bottom and middle. The pattern of haptic sensation changes as the object is held in different orientations. This prototype raises the question of how accurately a user could learn to estimate the orientation of the object from the 3 localized pulses.

Although only one *tone()* command can be used at a time, and it blocks all other routines, the 3 simultaneous pulse trains can be programmed using the *millis()* function that allows the tracking of time in the control loop. However the scheduling of the control loop varies with CPU load. At orientations with higher pulse rates in all 3 directions the pulses becomes irregular and glitchy due to the technical limitations of the Arduino to sustain the timing of the control loop at real-time rates. The trials also showed practical problems with the screwcap attachment. After several openings the wires from the battery became tangled and eventually detached, whilst multiple removals of the Ardweeny from the cavity detached the solder connections to the vibrators.

Jetsam demonstrated that haptic vibrations could also be an interface to Smart Things. This experiment also reiterated technical issues around the blocking behavior of the *tone()* command that effect multichannel sonic and haptic feedback on the Arduino.

The next iteration, called Lagan, is crafted from a cuttlefish backbone which is a natural material that floats and provides shock-proofing, with the benefit that it is lighter than wood. Two half shells were hollowed out and lacquered with polyester to seal them, as shown in Fig. 8.6. The problems of twisted connections in Jetsam led us to reuse the sealed gasket approach from Flotsam. This time we trialed an Arduino Nano that has a 43 × 19 × 10 mm footprint similar to the Ardweeny, with the added advantage of a lower input voltage of 6 V which allows a further reduction in size through the use of a flat Li-ion battery.

The sonification was designed to have a sea wave-like sound to match the organic shape and material. The sound was synthesized by varying the amplitude of white noise loaded into a cycling buffer and output to the PWM analog audio output. The need to synthesize a continuous sound without blocking data acquisition and control routines required the default timer behavior to be overridden. The variation in the acceleration in 3 directions is sonified by continuous variation in the amplitude of the synthesized sea-wave sound. The random number generator

Fig. 8.6 Smart Lagan

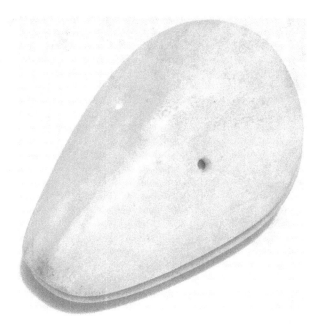

on the Arduino was too slow to interactively synthesize 3 channels of noise, so the number of channels was reduced to two. A pair of speakers were constructed on the object by gluing piezo buzzers against holes drilled through the housing that were sealed with thin flexible plastic to make them waterproof. The sound is audible but an amplifier could improve the dynamic range.

The development of a continuous sound for the Lagan prototype required changes to the low level workings of the Arduino software. This development adds a new capability to synthesize continuous sonic feedback without blocking the data acquisition and control. Building on this technical breakthrough we developed a continuous noise synthesizer that can be interactively varied in amplitude in response to a continuous stream of real-time data from a sensor.

Flotsam, Jetsam and Lagan were curated as an installation for the Conference on New Interfaces for Musical Expression in Sydney in 2010 (Barrass and Barrass 2010), shown in Fig. 8.7.

These prototypes provided insights and understanding of the design space of embedded sonifications as interfaces to Smart Things. We found that the Arduino microprocessor can be an alternative to consumer mobile devices as a platform for sonification in outdoor activities. However the *tone*() command limits the range of sounds to beeps and sine tones, hindering the development of more than very simple interfaces. Through these experiments it became clear that a naive approach to programming audio on Arduino would not provide satisfactory real-time performance. However, we were able to take over the timers on the Arduino to program a custom synthesis algorithm opening up the potential to design more complex sonifications.

Fig. 8.7 Flotsam, jetsam and lagan at NIME 2010

8.5 Mozzi: An Embeddedable Sonic Interface to Smart Things

The low level modifications to the Arduino required to synthesise the sea-sound sonification for Lagan provide a foundation for other sonic metaphors and synthesis algorithms. This led to the idea to develop a general purpose synthesis library to enable a much wider and richer range of sonifications. Mozzi is an open source software library that enables the Arduino microprocessor to generate complex and interesting sounds using familiar synthesis units including oscillators, samples, delays, filters and envelopes. Mozzi is modular and can be used to construct many different sounds and instruments. The library is designed to be flexible and easy to use, while also aiming to use the processor efficiently, which is one of the hurdles preventing this kind of project from succeeding until now. To give an idea of Mozzi's ability, one of the example sketches which comes with the library demonstrates fourteen audio oscillators playing simultaneously while also receiving real-time control data from light and temperature sensors without blocking.

Mozzi has the following features:

- 16384 Hz audio sample rate with almost-9-bit STANDARD and 14 bit HIFI output modes.
- Variable control rate from 64 Hz upwards.
- Familiar audio and control units including oscillators, samples, filters, envelopes, delays and interpolation.

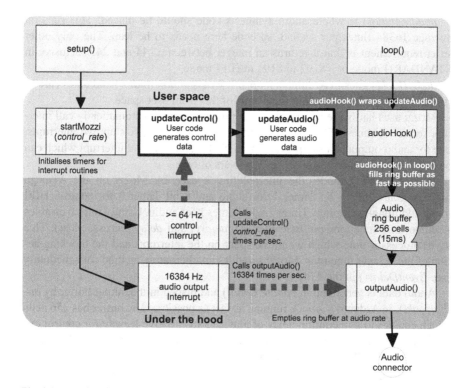

Fig. 8.8 Mozzi architecture

- Modules providing fast asynchronous analog to digital conversion, fixed point arithmetic and other cpu-efficient utilities to help keep audio running smoothly.
- Readymade wave tables and scripts to convert sound files or generate custom tables for Mozzi.
- More than 30 example sketches demonstrating basic use.
- Comprehensive API documentation.
- Mozzi is open source software and easy to extend or adapt for specific applications.

Mozzi inherits the concepts of separate audio and control rate processes directly from Csound (Vercoe 1993) and Pure Data (Puckette 1996). The interface between Mozzi and the Arduino environment consists of four main functions *startMozzi()*, *updateAudio()*, *updateControl()* and *audioHook()*, shown in Fig. 8.8. All four are required for a Mozzi sketch to compile.

startMozzi(control_rate) goes in Arduino's setup(). It starts the control and audio output timers, given the requested control rate in Hz as a parameter.

updateControl() is where any analog input sensing code should be placed and relatively slow changes such as LFO's or frequency changes can be performed. An example of this is shown in section 5.2.

updateAudio() is where audio synthesis code should be placed. This runs on average 16384 times per second, so code here needs to be lean. The only other strict requirement is that it returns an integer between -244 and 243 inclusive in STANDARD mode or -8192 to 8191 in HIFI mode.

audioHook() goes in Arduino's *loop()*. It wraps *updateAudio()* and takes care of filling the output buffer, hiding the details of this from user space.

Mozzi uses hardware interrupts on the processor which automatically call interrupt service routines (ISR) at regular intervals. *startMozzi()* sets up two interrupts, one for audio output at a sample rate of 16384 Hz and a control interrupt which can be set by the user at 64 Hz or more, in powers of two.

In STANDARD mode, the internal timers used by Mozzi on the ATmega processors are the 16 bit Timer 1 for audio and 8 bit Timer 0 for control. HIFI mode additionally employs Timer 2 with Timer 1 for audio. Using Timer 0 disables Arduino time functions *millis()*, *micros()*, *delay()* and *delayMicroseconds()*. This saves processor time which would be spent on the interrupts and the blocking action of the delay() functions. Mozzi provides an alternative method for scheduling (see *EventDelay()* in the API).

Audio data is generated in *updateAudio()* and placed in the output buffer by *audioHook()*, in Arduino's *loop()*, running as fast as possible. The buffer has 256 cells which equates to a maximum delay of about 15 ms, to give leeway for control operations without interrupting audio output. The buffer is emptied behind the scenes by the regular 16384 Hz audio interrupt.

Mozzi employs pulse wave modulation (PWM) for audio output. This allows a single Arduino pin to be allocated for output, requiring minimal external components. Depending on the application, the output signal may be adequate as it is. Passive filter designs to reduce aliasing and PWM carrier frequency noise are available on the Mozzi website if required. Mozzi has an option to process audio input. The incoming sound is sampled in the audio ISR and stored in a buffer where it can be accessed with getAudioInput() in updateAudio().

8.5.1 Application Programming Interface (API)

Mozzi has a growing collection of classes for synthesis, modules containing useful functions, commonly used wave tables, and sampled sound tables. Mozzi includes Python scripts to convert raw audio files and templates which can be used to generate other custom tables.

Descriptions of the classes currently available are shown in Table 8.1. Modules are described in Table 8.2. Comprehensive documentation of the library is available online http://sensorium.github.com/Mozzi/

Bare bones example: playing a sine wave This section explains a minimal Mozzi sketch step by step. The sketch plays a sine wave at a specified frequency. Although there are abundant instances online of Arduino sketches performing this task, this example illustrates the structure and gist of a bare-bones Mozzi sketch. It does not assume much previous experience with Arduino programming.

Table 8.1 The current collection of Mozzi classes, with descriptions and the update rates of each

Class	Description	Audio rate	Control rate
Oscillators			
Oscil	Plays a wavetable, cycling	Yes	Yes
Sample	Plays a wavetable, with extra controls	Yes	Yes
Phasor	Generates a high resolution ramp	Yes	Yes
Filters			
StateVariable	12db resonant lp, hp, bp and notch	Yes	–
LowPassFilter	Resonant low pass filter	Yes	–
Smooth	IIR low pass filter	–	Yes
Envelope generators			
ADSR	Simple ADSR envelope generator	–	Yes
Ead	Exponential attack decay envelope	Yes	Yes
Delays			
AudioDelay	Felay for comb filter, flange, chorus and	Yes	–
AudioDelayFeedback	Slapback	Yes	–
EventDelay	Audio delay with feedback	–	Yes
ReverbTank	Non-blocking replacement for Arduino's delay()	Yes	–
	Simple recirculating reverb		
Synthesis/Distortion			
WavePacket	Overlapping streams of windowed sine wave grains	Yes	–
WaveShaper	Maps input values through a table to output	Yes	Yes
Interpolation			
Line	Efficient linear interpolator	Yes	Yes
Portamento	Simple portamento for note-based applications	–	Yes

Table 8.2 Modules and descriptions

Module	Description
Fixmath	Fixed point fractional number types and conversion routines
Analog	Fast and non-blocking functions for reading analog input asynchronously, speeding up or replacing Arduino analogRead()
Midi	Midi note number to frequency conversions at various resolutions
Random	Fast pseudo random number generator functions
Core	Core definitions and functions

First include MozziGuts.h. This is always required, as are headers for any other Mozzi classes, modules or tables used in the sketch. In this case an oscillator will be used, and a wavetable for the oscillator to play:

```
#include <MozziGuts.h>
#include <Oscil.h>
#include <tables/sin2048_int8.h>
```

The oscillator needs to be instantiated using literal numeric values as template parameters (inside the <> brackets). This allows the compiler to do some of the Oscil's internal calculations at compile time instead of slowing down execution by repeating the same operations over and over while the program runs. An oscillator is declared as follows:

```
Oscil <table_size, update_rate> name(table_data);
```

The table size must be a power of two, typically at least 256 cells and preferably longer for lower aliasing noise. This Oscil will be operating as an audio generator. AUDIO_RATE is internally defined by Mozzi, and provided here so the Oscil can calculate frequencies in relation to how often it is updated. The table_data is an array of byte sized cells contained in the table file included at the top of the sketch.

The audio sine tone oscillator is created like this:

```
Oscil <2048, AUDIO_RATE> aSin(SIN_DATA);
```

The control rate, like the audio rate, must be a literal number and power of two to enable fast internal calculations. It is not necessary to define it as follows, but it helps to keep programs legible and simple to modify.

```
#define CONTROL_RATE 128
```

Now to the program functions. In Arduino's *setup()* routine goes:

```
startMozzi(CONTROL_RATE);
```

This sets up one timer to call *updateControl()* at the rate chosen and another timer which works behind the scenes to send audio samples to the output pin at the fixed rate of 16384 Hz.

The oscillator frequency can be set in a range of ways, but in this case it will be with an unsigned integer as follows:

```
aSin.setFreq(440u);
```

Now Arduino's *setup()* function looks like this:

```
void setup(){
  startMozzi(CONTROL_RATE);
  aSin.setFreq(440u);
}
```

The next parts of the sketch are *updateControl()* and *updateAudio()*, which are both required. In this example the frequency has already been set and the oscillator just needs to be run in *updateAudio()*, using the *Oscil::next()* method which returns a signed 8 bit value from the oscillator's wavetable. The int return value of *updateAudio()* must be in the range −244 to 243.

```
void updateControl (){
  // no controls being changed
}

int updateAudio (){
  return aSin.next();
}
```

Finally, audioHook() goes in Arduino's loop().

```
void loop(){
  audioHook ();
}
```

This is where the sound actually gets synthesised, running as fast as possible to fill the output buffer which gets steadily emptied at Mozzi's audio rate. For this reason, it's best to avoid placing any other code in *loop()*.

It's important to design a sketch with efficiency in mind in terms of what can be processed in *updateAudio()*, *updateControl()* and *setup()*. Keep *updateAudio()* lean, put slow changing values in *updateControl()*, and pre-calculate as much as possible in *setup()*. Control values which directly modify audio synthesis can be efficiently interpolated with a *Line()* object in *updateAudio()* if necessary.

The whole sketch is shown in Program 1.

Program 1. Playing a sine wave at 440 Hz.

```
#include <MozziGuts.h>
#include <Oscil.h>
#include <tables/sin2048_int8.h>

#define CONTROL_RATE 128
Oscil <2048, AUDIO_RATE> aSin(SIN_DATA);

void setup(){
aSin.setFreq(440u);
startMozzi(CONTROL_RATE);
}

void updateControl(){
}

int updateAudio(){
return aSin.next();
}

void loop(){
audioHook();
}
```

Vibrato can be added to the sketch by periodically changing the frequency of the audio wave with a low frequency oscillator. The new oscillator can use the same wave table but this time it is instantiated to update at control rate. The naming convention of using a prefix of k for control and a for audio rate units is a personal mnemonic, influenced by Csound.

```
Oscil <2048, CONTROL_RATE> kVib(SIN_DATA);
```

This time the frequency can be set with a floating point value:

```
kVib.setFreq(6.5f);
```

Now, using variables for depth and centre frequency, the vibrato oscillator can modulate the frequency of the audio oscillator in *updateControl()*. kVib.next() returns a signed byte between − 128 to 127 from the wave table, so depth has to be set proportionately.

```
void updateControl (){
  float vibrato = depth * kVib.next();
  aSin.setFreq(centre_freq+vibrato);
}
```

The modified sketch complete with vibrato is listed in Program 2.

Program 2. Playing a sine wave at 440 Hz with vibrato.

```
#include <MozziGuts.h>
#include <Oscil.h>
#include <tables/sin2048_int8.h>

#define CONTROL_RATE 128
Oscil <2048, AUDIO_RATE> aSin(SIN_DATA);
Oscil <2048, CONTROL_RATE> kVib(SIN_DATA);

float centre_freq = 440.0;
float depth = 0.25;

void setup(){
kVib.setFreq(6.5f);
startMozzi(CONTROL_RATE);
}

void updateControl(){
float vibrato = depth * kVib.next();
aSin.setFreq(centre_freq+vibrato);
}

int updateAudio(){
return aSin.next();
}

void loop(){
audioHook();
}
```

While this example uses floating point numbers, it is best to avoid their use for intensive audio code which needs to run fast, especially in updateAudio(). When the speed of integer maths is required along with fractional precision, it is better to use fixed point fractional arithmetic. The mozzi_fixmath module has number types and conversion functions which assist in keeping track of precision through complex calculations.

The Mozzi software was developed over a two year period. Many optimisation problems existed which had to be teased out one by one. The project has found solutions to the problems of affordable and easily embedded audio synthesis and has broken out of the sample-playback, single wave beeping paradigm widely accepted as embedded audio to date. This opens the way to increased creative uses of the Arduino and other compatible platforms. The first release of the library, which is called Mozzi, was made available on Github in June 2012. The range of potential applications has yet to be explored, however some examples which have appeared so far include:

A musical fruit fly experiment for a science fair at The Edge in the State Library of Queensland. Kinetic and electronic artists Clinton Freeman, Michael Candy, Daniel Flood and Mick Byrne worked with Dr Caroline Hauxwell from the Queensland

University of Technology to produce an interactive installation where people could play chords which represented the different resistances of a group of pieces of fruit infested with fruit flies. According to the project documentation, resistance is sometimes used as a measure of fruit quality (Freeman 2013).

B.O.M.B.-Beat Of Magic Box-, a palm-sized interactive musical device by Yoshihito Nakanishi, designed for cooperative performance between novice participants. The devices communicate wirelessly and produce related evolving harmonic and rhythmic sequences depending on how they are handled (Nakanishi 2011).

There are several MIDI-based synthesisers using Mozzi as a synthesis engine. One example is ^ [xor] synth by Václav Peloušek, founder of the Standuino handmade electronic music project, with six voice polyphony, velocity sensitivity, envelopes, selectable wavetables, modulation and bit-logic distortions (Peloušek 2013). Others include Arduino Mozzi synthesizer vX.0 by e-licktronic, a mono synth with selectable wavetables, LFO and resonant filtering (e-licktronic 2013), and the ironically humorous FM-based CheapSynth constructed and played by Dave Green and Dave Pape in a band called Fakebit Polytechnic (Fakebit Polytechnic 2013).

Mozzi is relatively young yet ripe for a community of open source development and practice to emerge around it. It's easy to write new classes and to construct composite instrument sketches. Feedback from educators has shown that the library is able to be used by children. There is the potential for porting the library to new Arduino platforms as they become available. As it is, there is already a long to-do list, including a variety of half-finished sound generators and instruments, and the ever-receding lure of creative work beyond the making of the tool.

Mozzi expands the possibilities for sonification and synthesis in new contexts away from expensive hardware, cables and power requirements. The low cost and accessibility of Mozzi synthesis on open source Arduino provides a way to create applications and compositions adapted to a wide range of localized conditions. We hope that Mozzi will contribute towards a future where Smart Things sound smart, through sonic interfaces that growl, purr and sing, rather than simply beep or tweet.

References

Barrass S (2013) ZiZi: the Affectionate Couch and the Interactive Affect Design Diagram, in Sonic Interaction Design, MIT Press, pp. 235–243

Barrass T (2013a)

Barrass S, Barrass T (2010) Flotsam, Jetsam and Lagan. In: Muller L (curator) Room based installations. Proceedings of the Conference on New Interfaces for Musical Expression, Sydney, 2010. http://www.educ.dab.uts.edu.au/nime/program.php#Installation, accessed 20 February 2014.

Barrass S, Barrass T (2013) Embedding Sonifications in Things. Proceedings of the International Conference on Auditory Display (ICAD 2013), Lodz, Poland.

Barrass S, Schaffert N, Barrass T (2010) Probing Preferences Between Six Interactive Sonifications Designed for Recreational Sports, Health and Fitness, Proceedings of ISon 2010, 3rd Interactive Sonification Workshop, KTH, Stockholm, Sweden, April 7, 2010.

Bradley MM, Lang PJ (2007) The International Affective Digitized Sounds, 2nd edn. IADS-2: Affective Ratings of Sounds and Instruction Manual. Technical report B-3. University of Florida, Gainesville

Fakebit Polytechnic (2013) CHEAPSYNTH   http://www.fakebitpolytechnic.com/cheapsynth-concept-safety-info/, accessed 20 February 2014.

Freeman C (2013) Untitled Sound Project Two. http://reprage.com/post/28654178439/untitled-sound-project-two. accessed 20 February 2014.

Hancock PA, Pepe AA, Murphy LL (2005) Hedonomics: the Power of Positive and Pleasurable Ergonomics. Ergonomics in Design: the Quarterly of Human Factors Applications, vol 13, Number 1. Winter 2005, pp. 8–14.

Hughes L, Saul S, Stuckey H (2003) Experimenta House of Tomorrow Exhibition, BlackBox, the Arts Centre, 100 St Kilda Road, Melbourne, 5 September to 3 October 2003, http://experimenta.org/events/2003/events2003.htm, accessed 20 February 2014.

Igoe T (2007) Making Thing Talk: using sensors, networks, and Arduino to see, hear, and feel your world. Maker Media, Inc.

e-licktronic (2013). e-licktronic Arduino Mozzi mono Synth v1.0. http://www.youtube.com/watch?v=wH-xWqpa9P8, accessed 20 February 2014.

Nakanishi, Y (2011). B.O.M.B. ver.2 (2011). http://yoshihito-nakanishi.com/works/device/b-o-m-b-ver-2/. accessed 20 February 2014.

Norman D (1988). The Design of Everyday Things. Basic Books, New York. ISBN 978-0-465-06710-7

Peloušek V (2013) ^ [xor] synth. http://www.standuino.eu/devices/instruments/xor. accessed 20 February 2014.

Puckette MS (1996) Pure Data. Proceedings, International Computer Music Conference, San Francisco, USA, 1996.

Vercoe B (1993) Csound—A Manual for the Audio Processing System and Supporting Programs with Tutorials, Media Lab, M.I.T., 1993.

Chapter 9
EEG-Based Brain-Computer Interface for Emotional Involvement in Games Through Music

Raffaella Folgieri, Mattia G. Bergomi and Simone Castellani

9.1 Introduction

Several studies promote the use of Brain Computer Interface (BCI) devices for gaming. By a commercial point of view, games represent a very lucrative sector, and gamers are often early adopters of new technologies, such as BCIs. Despite the many attempts to create BCI-based game interfaces, too few BCI applications for entertainment are really effective, not due to the difficulties on the computer side in signal interpretation, but rather on the users' side in focusing on their imagination of movements to control games' characters.

Some BCIs, such as the Emotiv Epoc (Emotive System Inc.), easily allow the myographic interface to generate game commands, but are hardly able to correctly translate the pure cerebral signals into actions or movements. Difficulties are mainly due to the fact that BCIs are currently less accurate than other game interfaces, and require several training sessions to be used. On the contrary, music entertainment applications seem to be more effective and require a short training on BCI devices.

In section two, we present a short review on implicit and explicit use of BCI in games and of main BCI commercial models. Section three focuses on our approach in enhancing gamers' emotional experience through music: we present the results of preliminary experiments performed to evaluate our approach in detecting users' mental states. We also present a prototype of a music entertainment tool developed to allow users to consciously create music by their brainwaves. Lastly, we shortly present the state-of-the-art of our research. In section four, we present our

R. Folgieri (✉)
DEMM, Dipartimento di Economia, Management e Metodi quantitativi,
Università degli Studi di Milano, Milano, Italy
e-mail: raffaella.folgieri@unimi.it

M. G. Bergomi
Dipartimento di Informatica, Università degli Studi di Milano, Milano, Italy

Ircam, Université Pierre et Marie Curie, Paris, France

S. Castellani
CdL Informatica per la Comunicazione, Università degli Studi di Milano, Milano, Italy

N. Lee, (ed.), *Digital Da Vinci,* DOI 10.1007/978-1-4939-0536-2_9,
© Springer Science+Business Media New York 2014

conclusions and our opinions about BCIs' promising application in entertainment, with particular focus on music.

9.2 Implicit and Explicit BCI in Games: A Short Review

There are several examples of games using brain activity to involve moving a cursor on the screen, or guiding the movements of an avatar in a virtual game through the imagination of the movements (Ko et al. 2009; Pineda et al. 2003; Lalor et al. 2005). With this aim, among the different BCIs proposed by commercial companies, two small-sized, inexpensive devices are currently largely in use in the scientific and in the entertainment communities: the Emotiv Epoc (Emotive System Inc.) and the Neurosky MindWave (NeuroSky Inc.). Despite of the lower number of signals detected by the Neurosky BCI (it does not register facial expressions as Emotiv does), it appears more suitable for entertainment applications, both for the lower-cost and for the easiness-to-wear. Several games use these two BCI devices as command interface. For example, some games such as "Sniper Elite" (Sniper Elite) or "Call of Duty" (Call of Duty) translate EEG signals collected by a BCI to determine how much the degree of attention affects the aiming or the accuracy of the player's weapon. In the games "Metal Gear Solid" (Metal Gear Solid), "Silent Hill" (Silent Hill) or "Resident Evil" (Resident Evil) the stress level can influence the behavior of non-player characters.

Most of BCI-based game interfaces adopt a two-class approach as a modality for movement control, realizing a so-called limited explicit interaction. Other games use signals registered by BCIs to modify the environment or the level of the game, realizing, in this case, the implicit approach. This latter is an interaction process not based on direct, explicit and voluntary user's action, but on the users' state in a particular context instead. In games, some examples are given by the games "Bacteria Hunt" (Mühl et al. 2010), in which alpha brain rhythms levels are related to the controllability of the player's avatar, or "AlphaWoW" (Plass-Oude Bos 2010), based on the game "World of Warcraft", in which the user's avatar can transform into an animal following the alpha brain rhythm activity.

We must recall that commercial BCI (Allison et al. 2007) devices consists in a simplification of the medical EEG equipment, communicating an EEG response to stimuli by WiFi connection, allowing people to feel relaxed, to reduce anxiety and to move freely in the experimental environment or in the game, acting as in the absence of the BCI devices. BCIs collect several cerebral frequency rhythms: the Alpha band (7–14 Hz), related to relaxed awareness, meditation, contemplation, etc.; the Beta band (14–30 Hz), associated with active thinking, active attention, solving concrete problems; the Delta band (3–7 Hz), frontally in adults, posteriorly in children with high amplitude waves, found during some continuous attention tasks (Kirmizialsan et al. 2006); the Theta band (4–7 Hz), usually related to emotional stress, such as frustration and disappointment; the Gamma band (30–80 Hz), generally related to cognitive processing of multi-sensorial signals. Both the two considered commercial BCIs collect all the listed signals and eye blink, but the Emotiv

Epoc also registers myographic signals, allowing one to also use facial expressions and head movements to interact with a game environment. On the other hand, the Neurosky MindWave provides two custom measures: "attention" and "meditation" meter values, which indicate respectively the user's level of mental focus and the level of relaxation. For their characteristics, the two considered devices could be efficiently used in games both implementing the explicit interaction and the implicit approach described above.

9.3   Enhancing Gamers' Emotional Experience Through Music

In our opinion, to make BCI an interesting interaction modality, we should focus on enhancing the user experience, considering that this device provides a direct detection of the user's mental state (implicit paradigm) and at the same time it could represent a new means of control (explicit paradigm). Having in mind the use of the Emotiv Epoc and the Neurosky MindWave in entertainment applications related to music, we performed several exploratory experiments to select the brain channels and rhythms to be considered, reducing the users' training time. The Neurosky MindWave particularly represents a challenge, due to the apparently lower potentiality compared to the Emotiv Epoc. In fact, while the Emotiv BCI device reads brainwaves data from fourteen sensors, the Mindwave provides just one sensor, positioned on the frontal area of the scalp. However, the Mindwave BCI is inexpensive and more comfortable for users, both for the easiness of positioning the device on the scalp, and because it uses a dry sensor, while Emotiv implements wet ones. With the aim to use a BCI as a mean to improve the game experience, we decided to focus our research effort in applying BCI devices in music entertainment applications and in adapting music to user's mental state in games. The first approach implements the BCIs explicit paradigm, while the second focuses on the implicit use of brain signals detected by a BCI device.

We consider the music field both a specific application in entertainment and a necessary part of emotional involvement of users in games and other kinds of entertainment applications. In fact, the application of BCI devices to entertainment by a music point of view covers mainly two perspectives:

* game sounds and music adaptation to users' mental state;
* music entertainment applications, such as, for example, music games or music production or composition engines.

In our experiments and in current works, we focus on these two points, at first performing some preliminary experiments with the aim to individuate how to detect users' mental state corresponding to sounds eliciting specific emotions; after testing the possibility to make people able to consciously produce single music notes; and finally creating game environments to test the possibility of influencing games' music following or contrasting the players' mental state to enhance their emotional experience and involve them more in the plot. In the following paragraphs, we show the results obtained within these three phases of our research.

9.3.1   Implicit Approach: Preliminary Experiments

With the aim to evaluate the possibility of using the two selected BCI devices, despite the reduced number of sensors, and to investigate if we could individuate a method to efficiently detect any user's mental state, we performed specific experiments necessary to access to the further steps in a feasibility-study phase of our research. We designed these experiments' protocol with the aim to test the reliability of both the Emotiv Epoc and the Neurosky MindWave.

Materials and Methods

We based the considered mental state/emotion labels on the 2D valence/arousal model originating from cognitive theory. In this model, valence is represented on the X axis from highly negative to highly positive, and arousal is on the y axis, from calming/soothing to excited/agitated (Russell 1980). This model has been used to determine the apparent mood of music in several works (Lu et al. 2006; Laurier et al. 2009).

To test the generalization ability of the chosen EEG features patterns and associated labels, we elicited physiological emotional responses using music stimuli and sound stimuli from International Affective Digitized Sounds (IADS) database[1]. We based our experimental choices on findings deriving from (Ansari-Asl et al. 2007; Tomarken et al. 1992; Müller et al. 1999; Keil et al. 2001; Schmidt and Trainor 2001), respectively individuating the minimum set of electrode to use in emotion detection, cerebral laterality, connection of emotions to gamma and alpha band and frontal EEG activity distinguished valence of the musical excerpts. Specifically, we designed the experiments according to (Schmidt and Trainor 2001), because the two dimensions of emotion mainly considered by researchers are valence and intensity, but there are really few models related to brain activity taking into account both dimensions. The model proposed in (Davidson et al. 1979) is focused on the asymmetrical frontal EEG activity as an index of the valence of emotion experienced, whereas the model presented in (Schmidt and Fox 1999) is based on the absolute frontal EEG activity, reflecting the intensity of emotional experience. In our experiments, we used the same approach adopted in (Schmidt and Trainor 2001) to determine if we could obtain similar results using the Emotiv Epoc and the Neurosky MindWave.

The two devices yet provide to remove artifacts from band power signals, so we analyzed data using, as in (Schmidt and Fox 1999), a discrete Fourier transform (DFT), with a Hanning window of 1s width and 50% overlap. In our case power (microvolts-squared) was derived from the DFT output in all the rhythms. On all the EEG data, we performed a natural log transformation to reduce skewness. Of course, while with the Emotiv Epoc we could evaluate the asymmetrical frontal EEG activities, thanks to the higher number of electrodes present on the device, we had not the same possibility with the Neurosky MindWave. For this reason, data

---

[1] http://csea.phhp.ufl.edu/media/iadsmessage.html.

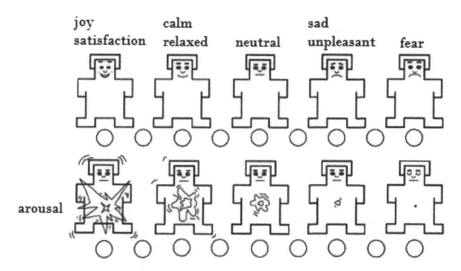

Fig. 9.1 The simplified version of the Self-Assessment Manikin (SAM) used in the experiments

have been collected for the both BCIs, but the valence measure has been performed directly only on data collected by the Emotiv Epoc. With data collected by the Neurosky MindWave we performed a comparison with the outcomes from the Emotiv data, obtaining an indirect measure of valence, corresponding to sound stimuli. On the contrary, the intensity measure has been performed on both BCIs.

Experiments Organization

The response to sounds and the obtained classification algorithm have been extensively evaluated in 60 right-handed participants in the experiments, equally distributed in age (20–49), gender and musical background, providing valence/arousal values as well as ensemble means and variance. Left-handed individuals were excluded because of possible differences in hemispheric specialization for emotion.

Each participant took the test separately, in a comfortable environment, to reduce variation influenced by external noises. Each EEG registration session had the duration of five minutes, during which subjects were resting, recommended to not close their eyes, and even not to speak or move while listening to the presented sounds. Between any couple of sounds, participants did not hear any sound for ten seconds.

We organized two experimental sessions: in the first brain rhythms Alpha, Beta, Delta, Theta and Gamma have been collected using the Emotiv Epoc; in the second we used the Neurosky MindWave. Each participant took part in both sessions.

Self-assessment of valence/arousal was therefore performed in the study by each participant and for each sound, using a simplified version of the Self-Assessment Manikin (See Fig. 9.1; Morris 1995). In this way, we could correctly identify the correspondence of hypothesized mental state response to each sound with the moods declared by the participants.

Table 9.1 Valence/hemisphere analysis (ANOVA) on all the bands

Rhythm	Valence		Hemisphere		Interaction between V and H	
	F(1,59)	p	F(1,59)	p	F(1,59)	p
Alpha	16.59	<0.0005	12.53	<0.0005	9.78	<0.002
Beta	15.76	<0.0005	11.62	<0.0005	8.67	<0.002
Delta	11.84	<0.004	10.65	<0.004	7.76	<0.004
Theta	11.52	<0.0005	10.57	<0.0005	8.53	<0.004
Gamma	15.54	<0.0005	11.52	<0.0005	8.87	<0.002

Results

Results showed that brain activity measured at the anterior part of the scalp distinguished the valence of musical emotions both using the Emotiv Epoc and the Neurosky MindWave. The activity increases during the presentation of positive-valence musical excerpts, while it decreases corresponding to negative ones. The overall frontal activation is related to the intensity of emotions elicited by music. In fact, it decreases correspondence to sound related to fear and increase for happiness or satisfaction. There is no difference in gender response.

On data collected with the Emotiv Epoc, we performed ANOVA between Valence (pleasant, unpleasant), Intensity (Intense, Calm), and Hemisphere (left, right) on ln(EEG power) separately for each brain rhythm and for frontal and parietal regions. The parietal analysis did not reveal significant variations. On the contrary, the analysis of EEG data from the frontal region showed a significant interaction among valence, intensity and hemisphere.

We performed ANOVA at first on valence and hemisphere, after on intensity and hemisphere, last on valence and intensity for all the rhythms.

As an example, in the following Table 9.1 we show a summary of the results obtained by ANOVA performed on valence/hemisphere analysis.

Positive-valence musical stimuli elicited greater left frontal EEG activity and less frontal EEG power in alpha and beta bands than in theta and gamma bands, while negative-valence musical stimuli elicited greater right frontal EEG activity and less frontal EEG power in theta and gamma bands than in the alpha and beta bands.

According to (Schmidt and Trainor 2001), the intensity/hemisphere analysis showed significant effects of intensity and hemisphere in all the bands, but did not reveal any interaction. We observed less overall frontal EEG power (i.e. more activity) corresponding to intense music stimuli compared to music stimuli eliciting less intense mental states.

Last, the valence/intensity analysis showed the effects of valence and intensity. As in intensity/hemisphere analysis, less overall frontal EEG power (i.e. more activity) corresponds to intense music stimuli compared to music stimuli eliciting less intense mental states.

We performed these measures directly on data collected by the Emotiv Epoc, while, to detect the same response on data collected with the Neurosky MindWave, we had to compare signals, due to the presence of a single sensor not allowing the hemisphere analysis. Despite of this, even with a lower precision, also with the Neurosky MindWave the experiments led us to well-individuate four mental states by EEG analysis, we related to involvement of players in games: fear, related to arousal due to stress; joy, corresponding to satisfactory; happiness, related to arousal due to relaxation; sad, corresponding to frustration.

We found that, both the Emotiv Epoc and the Neurosky MindWave have the sufficient number of sensors needed for our objectives. Therefore, the Emotiv Epoc showed a greater precision in detecting features individuating the selected users' mental states (89 vs. 81 %), probably due to the direct measure of the valence and hemisphere. Moreover, users prefer the Neurosky MindWave. In fact, for its wearability, users could wear it for a long time, without feeling awkward. On the contrary, when we tested the Emotiv Epoc for a long time, users tended to lose their attention and feel frustrated for the uncomfortable shape of the device. We detected, in fact, an increase in Theta band activities and a corresponding decrease in the other rhythm's power. Considering these results, we decided to use indifferently the Neurosky Mind-Wave in future experiments to refine the approach, sacrificing a greater precision for a long-lasting data detection without decay of the involvement level of the users.

9.3.2   Implicit Approach: Quantitative and Qualitative Analysis of Perception of the Emotional Interaction Between Visual and Audio Stimuli

The aim of this work has been to investigate how brain reacts to the perception of stimuli belonging to different sensorial areas. The focus of this experiment is on the β rhythm since it represents the activation state of a subject. Therefore, what we measure is the capability of an audio stimulus to change the preexisting activation state of a subject boosting or weakening it.

In particular, by inducing a state through a visual stimulus, we analyzed how the β-wave production changed after an audio stimulus emotionally coherent to the projected image is proposed to the subject, to finish with a new audio stimulation with an opposite emotive character respect to the visual one.

The first problem to face is the stimuli choice and pairing, which is at the same time, a preliminary problem, and a core step to the realization of the experiment. The concepts of emotional coherence and opposition for visual and audio stimuli are defined, and an automated algorithm for the paring of correlated and opposite stimuli is suggested in paragraph "An Algorithm to Find Correlated and Opposite Stimuli".

The experiment is built on a starting set of 7 3-uples of stimuli, each vector composed by a visual stimulus and two audio stimuli in emotive coherence and opposition to the visual one. The outline of the experiment is described in paragraph "Experimental Outline".

The preprocessing techniques applied to the signal acquired during the experiment are described in paragraph "Signals Preprocessing". After a brief introduction to the most common artifacts and their typical patterns observable in the EEGs, the ICA method is described and applied to the signals collected by 6 of the 14 channels of the Emotiv Epoch.

The preprocessing phase ends with the application of a Čebyšev band-pass filter to the cleaned signal, to highlight the β-wave's frequency bandwidth.

In paragraph "Analysis and Results", the elementary statistical methods that we used to compute the activation of our subjects are shown, and analyzed applying the Q-transform to the averaged signal we obtained to for each experimental set.

An Algorithm to Find Correlated and Opposite Stimuli

Since the aim of this work is to give a quantitative and qualitative analysis of the perception of correlated, and opposite audio and video stimuli, a fundamental step is to choose them properly.

This choice implies a certain number of intermediate steps in which some assumptions have to be made, some stimuli's feature have to be extracted, and finally, the definitions of emotionally correlated and opposite stimuli have to worked out. All these tasks are resumed in the following list:

1. Give a 2-dimensional representation of International Affective Picture System (IAPS) and International Affective Digitized Sounds (IADS), which is a geometrical representation of the standard dimensional valence/arousal model and organize data thanks to a square grid.
2. Find a space in which IADS and IAPS can be embedded and interact.
3. Define audio and visual stimuli correlation and opposition.

At this point, the creation of an algorithm that is able to choose the 3 components vector of stimuli we are looking for, is straight-forward.

In the following paragraphs we shall analyze each passage. In paragraph A.3 the flow diagram associated to the algorithm is shown.

A Geometrical 2-Dimensional Representation of IAPS and IADS

This first step is trivial: in both IAPS and IADS database stimuli are classified thanks to their valence and arousal values. These 2 values are exactly the Cartesian coordinate of each stimulus in R^2. In Fig. 9.2, the two point clouds representing respectively the IAPS and IADS database are shown. We placed a square grid, which divide each cloud in valence/arousal classes. In practice this is equivalent to a clusterization of data contained in each database. The rule to associate stimuli can be written as an equivalence relation as follows.

Let α_1 and α_2 be two stimuli in IADS. They can be written as $\alpha_1 = (a_1; v_1)$ and $\alpha_2 = (a_2; v_2)$.

The two stimuli are in relation, $\alpha_1 \sim \alpha_2$ if and only if

$$\lfloor a_1 \rfloor = \lfloor a_2 \rfloor \, and \lfloor v_1 \rfloor = \lfloor v_2 \rfloor,$$

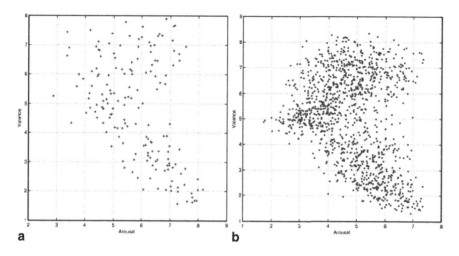

Fig. 9.2 IADS (**a**) and IAPS (**b**) as point cloud in $\mathbb{R}^2$

where $\lfloor \cdot \rfloor$ denotes the floor function, thus if $x \in \mathbb{R}$

$$\lfloor x \rfloor = \max\{k \in Z : k \leq x\}.$$

IADS ∪ IAPS in $\mathbb{R}^3$

The second step allows us to visualize at the same time, both the IADS and IAPS databases in the same geometrical space. Since we represented them as planes it is natural to embed them in $\mathbb{R}^3$ as two parallel planes. See Fig. 9.3.

The plane nearest to the observer represents the IADS' point cloud, while the further one is associated to the IAPS database.

The great advantage of this kind of representation is that we can define two simple functions to project points belonging to one plane to the other and vice versa. Consider a stimulus α belonging to the point cloud associated to the IADS database, embedded in the 3- dimensional space, then its coordinate are of the form (a, v, 0), if we want to project this point on the plane where the IAPS point cloud lies, suffice to map α in its projection (a, v, 10). Formally we can write the following

Definition Let $\alpha \in \mathbb{R}^3$, $\alpha = (a_\alpha, v_\alpha, 0)$, we define $P: \mathbb{R}^3 \to \mathbb{R}^3$ such that

$$P(\alpha) = (a_\alpha, v_\alpha, 10).$$

Symmetrically, we can choose a point $\iota = (a_\iota, v_\iota, 10)$ and map it through a function Q to the plane where the IADS' point cloud lies:

$$Q(\iota) = (a_\iota, v_\iota, 10).$$

The functions P and Q allow us to find the shadow of a certain audio stimulus on the plane where pictures lie and vice versa. The idea is to choose a feature proper of

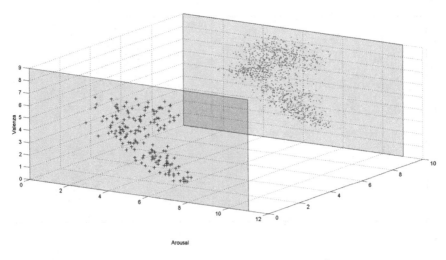

Fig. 9.3 A geometrical representation of IADS and IAPS database in $\mathbb{R}^3$

these stimuli that can be used either to choose if they are either coherent or opposite on an emotional point of view.

Emotional Comparison Between Audio and Visual Stimuli

Sometimes the easiest solution to a problem works properly, our assumption to compare stimuli belonging to such different context is that they both have been classified thanks to a statistical analysis of an experiment, thus it was possible that the position of a certain audio stimulus identified by its coordinate in the IADS' plane, could be associated at least with a certain set of images in the corresponding position on their plane.

Let's consider an audio stimulus α. We tried to find a set of corresponding visual stimuli by simply mapping it to the IAPS' plane through the function P defined above. $P(\alpha)$ belongs to a certain area of the grid traced on the IAPS' plane, see paragraph "A Geometrical 2-Dimensional Representation of IAPS and IADS".

To clarify this passage, take a look to Fig. 9.4, without loss of generality, assume that α belongs to the class of arousal 7 and valence equal to 6. The number you read in the grey cell corresponding to the grid square (6; 7) is the number of audio stimuli contained in that class, thus we can choose an audio stimulus α in that class since it is non-empty. The map P projects α to the IAPS' plane in a position corresponding to its valence/arousal coordinates, thus the $P(\alpha)$ belongs to the same square (6; 7) in the IAPS' plane. Thanks to the Fig. 9.4 we can see that this region of the IAPS' point cloud is non-empty. We found the set of visual stimuli associated to the starting audio α.

Since the aim of this selection is to feed an experimental protocol, we have to choose only one of the stimuli belonging to the visual coherent set we found. To do

IADS		Valence						
		1	2	3	4	5	6	7
Arousal	1							
	2					1		
	3				1	3	2	1
	4			1	14	13	6	4
	5		3	5	9	7	12	3
	6		11	19	2	6	7	8
	7	8	11	1			5	3
	8		1					

IAPS		Valence							
		1	2	3	4	5	6	7	8
Arousal	1				2				
	2				33	24	8	2	
	3		1	12	56	98	42	29	1
	4	1	25	45	44	66	82	57	12
	5	17	75	65	28	25	70	51	3
	6	32	57	20	9	12	38	22	1
	7	9	2					2	4

a   b

Fig. 9.4 Number of stimuli in IADS (**a**) e IAPS (**b**) for each unitary class of valence and arousal

that, we decided to find the visual stimuli which is the nearest to the projection of α. In symbols the picture $\overline{\imath}$ associated to α is given by

$$\overline{\imath} = \min_{\imath \in IAPS} d(P(\alpha), \imath),$$

where d is a distance.

Remark We tried to use different method to compute the distance between stimuli, such us the taxicab norm, the L^p norm ($p \geq 1$) and Hausdorff distance, however the results we obtained was exactly the same as the ones obtained with the Euclidean distance.

So far, what we have is the starting stimulus α and the picture $\imath$ which is correlated to the audio sample.

A new audio stimulus in opposition to $\imath$ has to be found. To do that our first idea has been to project $\imath$ on the IADS' plane and then find the stimulus which is at the maximum distance from $Q(\imath)$.

It suffices to take a look to Fig. 9.3 to observe that the points of IAPS have a higher density than the ones in IADS. Thus, it surely happens that a large set of pictures will be associated to the same audio stimulus.

To avoid this kind of problem firstly, we decided to come back to a problem of minimization, so given $Q(\imath)$ we took the point symmetrical to the projection respect to the line depicted in Fig. 9.5, let's call it $Q(\imath)$. After this procedure the audio stimulus in opposition to the picture $\imath$ is the audio stimulus that is nearest to $Q(\imath)$. Thus the problem is symmetrical respect to the one discussed above to find the picture correlated to an audio sample.

In spite of this strategy it could happen that more than one picture is associated to an opposite audio stimulus, the reason is the same density argument given before. This kind of situation is not suitable to the experiment: each stimulus has to be proposed to the subject only one time. Thus, if a certain audio has already been tagged as the audio in opposition to another image, the algorithm temporarily deletes that audio point from the cloud and finds a new minimum. This step is iterated until each 3-uple composed by the starting audio, the visual stimulus correlated and the audio stimulus in opposition, is unique. Clearly, since the datasets we are working with are finite, it is not possible to generate an infinite number of unique vectors.

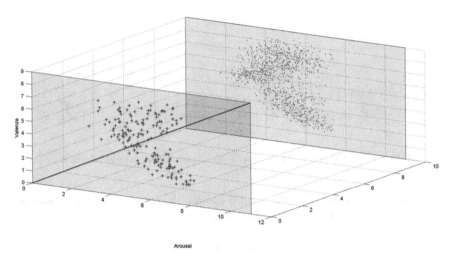

Fig. 9.5 Axis of symmetry of the IADS plane

However, in our experiment we use a set of 7 starting audio, thus the uniqueness of the vectors is guaranteed.

What has been interesting is that thanks to the empirical nature of the data contained in IADS and IAPS, and the small cardinality of the starting set of audio stimuli, even if we have automated the choice of the stimuli we are not renouncing to the symbolical meaning of the stimuli. For instance, what happens is that starting from an audio sample of children playing, the image associated to the stimulus is a park. Furthermore considering the awful noise of a dental burr, an image in which a surgical intervention is shown explicitly is the correlated stimulus to the audio, and the sound of a waterfall is the opposite one.

Experimental Outline

Thanks to the algorithm described in Fig. 9.6, we selected seven triplets that are organized to form 7 stimulation-set, that will be noted as $set(i)$ for $i \in \{1,\ldots,7\}$, which is presented to the subjects in this order:

1. picture ι;
2. picture with coherent audio (ι, α);
3. picture with opposite audio (ι, α').

The structure of a generic $set(i)$ is shown in Fig. 9.7 and the stimulation protocol we adopted is shown in Fig. 9.8; it can be noticed how the protocol iterates itself from the second block since the first five seconds of black screen are used to relax the subject before the experiment starts.

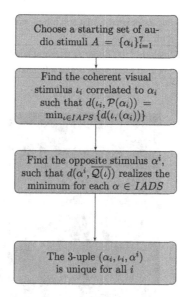

Fig. 9.6 Algorithm for axis of symmetry of the IADS plane

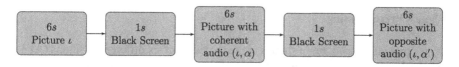

Fig. 9.7 A generic set(i)

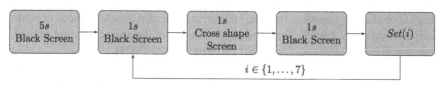

Fig. 9.8 Stimuli presentation protocol

Each stimulus is presented for 6 s. Between two consecutive stimuli 1 s of black screen is shown. In agreement with (Petrantonakis and Hadjileontiadis 2012), before each stimulation-set, a cross shape in the middle of the screen is shown for 1 s to attract the sight of the subject.

Table 9.2 shows the values of valence and arousal of each selected stimulus. Stimulation is organized in such a way that maximum opposition between positive picture and negative audio stimuli are projected at the end of the experiment; this is to avoid possible effects like emotional cover-up, due to a presentation of a more activating stimulus than a less one (Petrantonakis and Hadjileontiadis 2012).

Table 9.2 Examples of notes and associated visual and motor stimuli

Note	Visual stimulus	Motor stimulus
A	Orange	Knock the first finger of the right hand on the thumb of the left hand
C	Blue	Knock all the fingers of the left hand on the palm of the right hand

Fig. 9.9 Artifacts in single channel EEG recorded from Neurosky

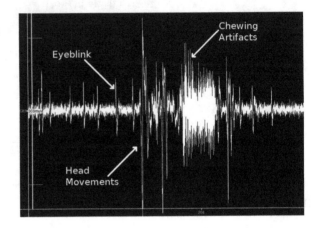

Signals Preprocessing

As we previously said, brain signals are recorded from six different positions (AF3, AF4, F3, F4, O1, O2) and they may contain several artifacts due to eye-blinks, eye-movements and other kind of extra-cortical signals such as heartbeats and muscle activities. Though these artifacts have characteristic patterns (see Fig. 9.9), they often share the same portion of the frequency spectrum with the brain rhythms; then it could be that the complete removal of artifacts will also remove a piece of information of the EEG signals. Thus, the entire removal of the artifacts during acquisition process is quite impossible. See (Murugappan et al. 2009).

There are many methods used to preprocess acquired signals and remove artifacts, some of them use digital signal processing (DSP) tools like adaptive filters in cascade (Garcés Correa et al. 2007), whereas others use a statistical approach. In (Dyana Szibbo et al. 2012) these two different classes of methods are compared on removing blink artifacts in a single channel EEG and this comparison reveals that they both work fine for the purpose.

In our experiment we preprocess signals with the Independent Component Analysis (ICA) (Scott et al. 1996) algorithm, which will be explained in paragraph "Independent Component Analysis".

The last method we used is a Čebyšev band-pass filter, that was applied to cut frequencies outside the band occupied by the β rhythms.

Fig. 9.10 Model of a room containing 3 audio sources and 3 microphones

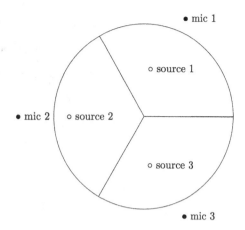

Independent Component Analysis

The ICA algorithm is typically used to solve the so-called "cocktail party problem", that is related to the blind source separation (BSS), where the voice of a person have to be recognized in a signal that is the mixture of several sources (Hyvärinen and Oja 2000).

In Fig. 9.10, a room in which three sources are generating audio signal and three microphones are recording them, is shown. What happens is that each microphone records a linear combination of the signal produced by the three sources. We give a practical example in Fig. 9.11: in the first row, the three sources are shown. In the second one, three different linear combinations of the signals are depicted, think about them as the signal recorded by each microphone of Fig. 9.10.

By applying the ICA algorithm, we obtain the third row of the figure that shows the independent components of the linear combinations. Though the amplitude and the order of the components are not the same as in original sources (first row of the figure), it could be noticed that they are rather the same. That's not all; thanks to its operation, ICA is widely used as a preprocessing and de-noising tool (bin Yunus 2012). If we assume that one of the three source signals (i.e., the middle one in Fig. 9.11a) is a noisy component, we can set it to zero and compute the inverse ICA algorithm on the remaining independent components (ICs) to remove it from the mixed signal, the result of this operation is shown in Fig. 9.11d.

In our study, this procedure has been applied to the six EEG signals recorded from the subject's scalp. These signals are assumed to be a linear combination of unknown independent components within the brain (Scott et al. 1996; Zhou and Gotman 2009). The ICs have been computed separately for each subject and then they have been visually analyzed. The process ended setting to zero the components identified as artifacts, to be subsequently deleted from the brain signals.

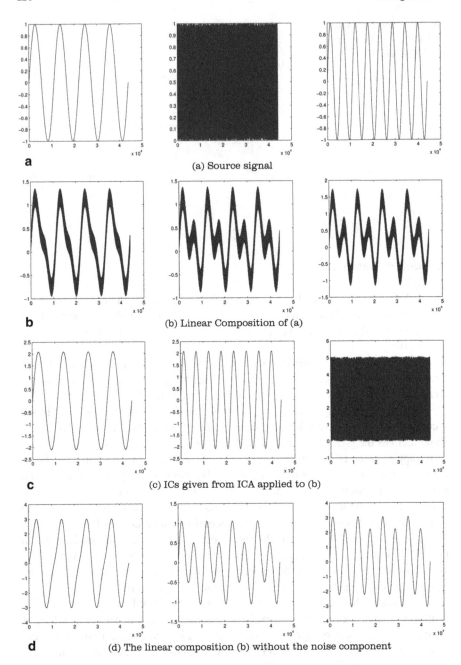

(a) Source signal

(b) Linear Composition of (a)

(c) ICs given from ICA applied to (b)

(d) The linear composition (b) without the noise component

Fig. 9.11 An example of the application of the ICA algorithm

Some Mathematical Details on ICA's Assumptions

In this paragraph we are going to go a little deeper in the analysis of the ICA method. Our main sources are (Comon 1994; Choi et al. 2005; Hyvärinen and Oja 2000; Lee 1998; Stone 2004).

Let's build a general example to understand what lies under the ICA.

Consider n linear combinations $\{x_1, ..., x_n\}$ of independent components $\{s_1, ..., s_n\}$, (for simplicity we maintained the notation of (Hyvärinen and Oja 2000), where you can find a deeper analysis of the ICA method). In symbols we have:

$$
\begin{aligned}
x_1 &= a_{1,1}s_1 + a_{1,2}s_2 + \cdots + a_{1,n}s_n \\
x_2 &= a_{2,1}s_1 + a_{2,2}s_2 + \cdots + a_{2,n}s_n \\
&\vdots \\
x_i &= a_{i,1}s_1 + a_{i,2}s_2 + \cdots + a_{i,n}s_n \\
&\vdots \\
x_n &= a_{n,1}s_1 + a_{n,2}s_2 + \cdots + a_{n,n}s_n.
\end{aligned}
$$

The time index t does not appear in the formula, since we are assuming that x_i and s_i, where $i \in \{1,...,n\}$, are random variables, and not proper time signals. We can think about $x_i(t)$ as a sample of the random variable x_i. It is common to represent the equations shown in the previous formula in matrix notation.

Set

$$
x = \begin{bmatrix} x_1 \\ x_2 \\ \vdots \\ x_n \end{bmatrix} \quad A = \begin{bmatrix} a_{1,1} & a_{1,2} & \cdots & a_{1,n} \\ a_{2,1} & a_{2,2} & \cdots & a_{2,n} \\ \vdots & \vdots & \ddots & \vdots \\ a_{n,1} & a_{n,2} & \cdots & a_{n,n} \end{bmatrix} \quad s = \begin{bmatrix} s_1 \\ s_2 \\ \vdots \\ s_n \end{bmatrix},
$$

Where x is the vector of the mixtures, A is called the mixing matrix and s is the vector of the independent components. The problem can be written as

$$\mathbf{x} = \mathbf{As}.$$

Denoting with $a_j, j \in \{1,...,n\}$ the columns of the matrix A, we have that

$$\mathbf{x} = \sum_{j=1}^{n} \mathbf{a}_j \mathbf{s}_j.$$

The model represented above describes the observed data as a mixture of independent components, that is exactly the ICA: this method allows us to determine the vector **s** and the matrix **A** given the set of observable data **x**.

Assumptions To efficiently model a problem with the ICA we have to assume that the components $\{s_1, ..., s_n\}$ are independent and have non-Gaussian distribu-

Fig. 9.12 The density function of the Laplacian distribution, which is a supergaussian distribution. The dashed line shows the Gaussian density

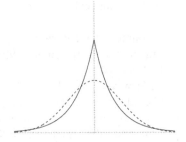

tions. Since the assumptions we made are elementary for the people who know the probability theory, but they surely sound exoteric for those who do not have this kind of background, here follows a simple definition of independent events (instead of random variables), and a description of non-Gaussian distributions to help the reader's intuition.

Definition Two events A and B are said to be independent if and only if the occurrence of A does not depend on the occurrence of B and vice versa. In symbols

$$P(A \cap B) = P(A)P(B).$$

Concerning non-Gaussian variables, suffice to observe that the joint density of two Gaussian variables is completely symmetric, thus it does not contain information on the direction of the columns of the mixing matrix, thus **A** cannot be estimated. In Fig. 9.12 one of the most common example of non-Gaussian distribution is depicted. See (Choi et al. 2005) for further details.

Analysis and Results

In this section a brief state of the art on physiological signal analysis is proposed, to introduce the methods used in this work. Then follows the presentation of the results we obtained.

Physiological Signal Analysis

Different methods have been adopted to extract features and classify emotions from electrophysiological signals: in (Murugappan et al. 2009) the emotions are classified with statistic features like standard deviation, variance, entropy, and amplitude of the EEG signals with a higher accuracy rate up to 90 %. In (Tseng et al. 2012) an index computed through the linear combination of the Mahalanobis distance between the different rhythms is used to classifying 4 users' cognitive states.

A features extraction algorithm called *asymmetrical spatial filter* that maximizes the difference between the variance of the EEG signal from the two different cortical hemispheres is proposed in (Huang et al. 2012). A similar approach is used in (Petrantonakis and Hadjileontiadis 2012) where the spatial asymmetry is combined with the *multidimensional directed information* approach to identify and find the epochs of EEG signals.

A neural network (NN) is used in (Ryu et al. 1998) to classify likable and unlikeable stimuli, the classification rule of the NN is based on the slop of the function that describes the trend of the asymmetric ratio in sub-bands of α and β rhythm.

Although (Gaetano Valenza et al. 2012) does not refer to EEG signals (but other peripheral physiological ones like electrocardiogram (ECG) and respiration activity) it compares the efficacy of a classifier based on two set of features: the first uses standard ones (like statistical values and features of time-frequency domain) and the second use non-linear methods to analyze the deterministic chaos and the recurrences of a state of the system through the Recurrence Plot (RP), treating signals like dynamical systems.

Results

The aim of our study is to observe the qualitative and quantitative variation of the β rhythms in the different stimulation sets, to define how the stimulation through the superposition of coherent and opposite stimuli can influence the brain's activation. The presence of these rhythms is related to the engagement of the subject and their observation could be useful to determine his mental state.

We looked for a variation trend of the subjects' activation that confirms the results given from the preliminary experiment for the implicit approach paradigm explained in Sect. 3.1.

To achieve that, a group of features that discriminates the level of the activation has to be found. Thus, we firstly epoched the time series signals, obtained from the application of the protocol of signal preprocessing (see Sect. C), in according with the duration of each stimulus, and after that we compute the average response of each subject like the point-to-point means of the six channels recorded on the scalp.

The root mean square (RMS) values of these average responses has been calculated on all the subjects, for each epoched data, to have the general averaged response of the population to a specific stimulation.

Constant Q Transform   For the analysis of our data we apply the constant Q transform (QT) to the averaged responses in each stimulation set.

The constant Q transform is not far from the standard discrete Fourier transform (DFT), with the difference between the two ways to compute the Fourier transform on a signal being that the frequency components computed via the DFT are separated by a constant Δ of frequencies and with a constant resolution. The Q transform works with a constant ratio of centre frequency to resolution.

The natural application of the QT is on musical signals, since it is equivalent to a 1/24-octave filter bank, so it has two different frequency components for each note

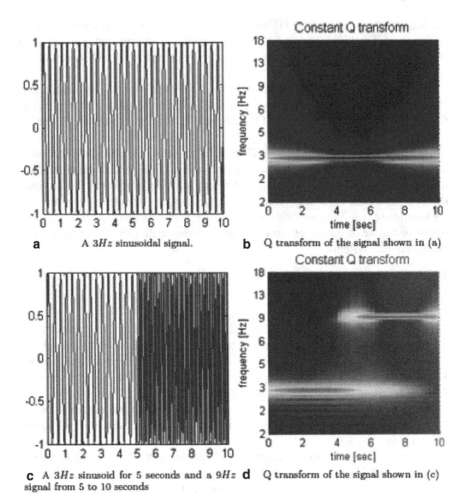

a   A 3*Hz* sinusoidal signal.   **b**   Q transform of the signal shown in (a)

c   A 3*Hz* sinusoid for 5 seconds and a 9*Hz*   **d**   Q transform of the signal shown in (c)
signal from 5 to 10 seconds

Fig. 9.13 Two examples of constant Q transform applied on sinusoidal signals

of an octave, this feature makes the Q transform able to distinguish among adjacent note played simultaneously. See (Brown 1991) for further details.

The reason that led us to choose this kind of computation is that the log frequency is more efficient than the linear frequency representation in the accomplishment of some tasks such as the visualization of the attack and decay of a certain signal. Our interest lies exactly in determining the state of activation and de-activation of a subject due either to a single stimulus or complex stimuli interaction.

Example Let's consider a 3 Hz sinusoidal signal of a duration of 10 s, see Fig. 9.13a. The constant Q transform of this signal is depicted in Fig. 9.13b, in which a red stripe represents exactly the frequency and the duration of the signal. In Fig. 9.13c, we analyzed a juxtaposition of a 3 Hz and a 9 Hz signal. Each one lasts 5 s. The

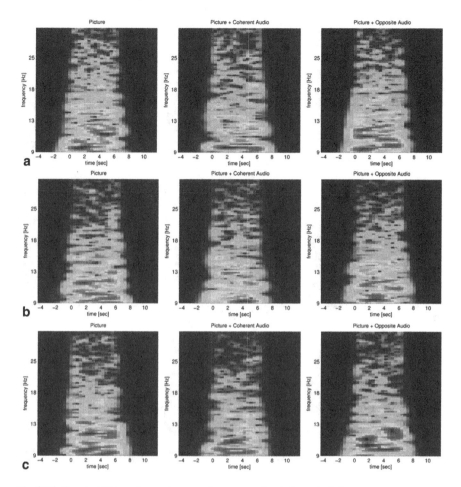

Fig. 9.14 Constant Q transform of averaged response stimulation sets 1–3

Q transform of this example (see Fig. 9.13d) reveals how we can even graphically understand the attack and release time of each signal.

The application of the QT to our signals is shown in Figs. 9.14 and 9.15. The first one shows the QT belonging to the stimulation sets from 1 up to 3 while the second one refers to the stimulation sets from 4 up to 7.

Table 9.3 shows schematically the patterns of the brain activation we found in the different kinds of stimulation. These patterns were obtained through the observation of the QT.

Low-valence Picture   On one hand the β rhythms production given by the stimulation through a low-valence picture can be increased by the presentation of a high-valence audio stimulus (see and compare the first and the third columns of the QT depicted in Fig. 9.14).

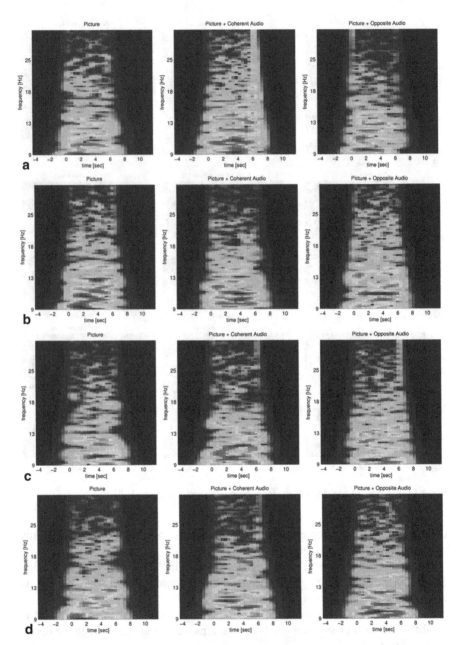

Fig. 9.15 Constant Q transform of averaged response stimulation sets 4–7

Table 9.3 Trends of the brain activity through the different kinds of stimulation

	Low-valence picture	High-valence picture
High-valence audio	Increasing of activity	Increasing of activity
Low-valence audio	Decreasing of activity	Variable trend

On the other hand, the stimulation given by the superposition of the same picture with a low-valence audio stimulus determines the decreasing of the production of β rhythms (see the QT depicted in the first two columns of the Fig. 9.14).

High-valence Picture The activation state induced by a high-valence picture can be increased thanks to the presentation of a high-valence audio stimulus (see and compare the first two columns of the QT depicted in Fig. 9.15).

In the last kind of stimulation i.e. the superposition of low-valence audio stimuli in opposition to pictures with high-valence level, we expected a greater activation with the increasing of the difference between the values of the stimuli's valence.

This is not what we observed. The brain activity patterns through the four stimulation sets have not a regular trend.

To better figure the homogeneity of the responses between the subjects in the different set, we apply a graphical methods borrowed from the statistics and chaos theory.

Recurrence Plot The recurrence plots (RPs) are largely used to study dynamical systems. To speak about RPs we have to give an intuitive definition of time series. It can be thought as a sequence of successive points in time, in the case of a signal the time series is given by the samples taken by the BCI interface to digitalize the EEG signal we want to process. A formal definition of time series and a deep treatise about them, dynamical systems and applications is provided in (Hamilton 1994).

To compute dynamical parameters from a time series, it is necessary that it be taken from an autonomous dynamical system, i.e. its evolution equation does not contain the time explicitly. In (Eckmann et al. 1987) recurrence plot was presented as a tool to test this assumption, and it is presented as a means to obtain *surprising and not easily obtainable information* from other models.

The RP is a graphical tool that works as follows. Consider a time series given by $\mathbf{x} = \{x(1),...,x(n)\}$. Let's $\overline{i} \in \{1,...,n\}$, for each component of the time series we define

$$R_{\overline{i},j} := \begin{cases} 1 \text{ if } d(x(\overline{i}),x(j)) \leq \varepsilon \\ 0 \text{ if } d(x(\overline{i})),x(j)) > \varepsilon \end{cases}$$

What we obtain is a $(n \times n)$ matrix $\mathbf{R} = R_{i,j}$, where $i,j \in \{1,...n\}$.

Example Set $x := \{1/2,2,5/2\}$, $\varepsilon=1$ and d$(\cdot,\cdot)$ to be the Euclidean distance, we have

$$R_{\frac{1}{2},\frac{1}{2}} = 1$$

$$R_{\frac{1}{2},2} = 0$$

$$R_{\frac{1}{2},\frac{5}{2}} = 0$$

$$R_{2,\frac{1}{2}} = 0$$

$$R_{2,2} = 1$$

$$R_{2,\frac{5}{2}} = 1$$

$$R_{\frac{5}{2},\frac{1}{2}} = 0$$

$$R_{\frac{5}{2},2} = 1$$

$$R_{\frac{5}{2},\frac{5}{2}} = 1.$$

Thus **R** is given by

$$R = \begin{bmatrix} 1 & 0 & 0 \\ 0 & 1 & 1 \\ 0 & 1 & 1 \end{bmatrix},$$

which is a symmetric matrix (i.e. $\mathbf{R} = \mathbf{R}^{\mathrm{T}}$) as we expected.

The RP representation is defined associating on a plane a black pixel at the coordinate (i, j) if $RR_{i,j} = 1$ and a white one, otherwise. See Fig. 9.16.

So, we want to investigate how much the responses between all the subjects are homogeneous in each of the 3 kinds of stimulation. We made 21 time series (one for each stimulus) that are the concatenation of a subsampling of the averaged responses of all the subjects. Then we applied the RP with two different values of the threshold to every time series. The result of the application of the RP to the signals belonging to the second stimulation set is depicted in Fig. 9.17.

To quantify the recurrence in our system, we calculated the Recurrence Rate (RR) that can be expressed as

$$RR = 100 * \left(\frac{1}{N^2} \sum_{i,j=1}^{N} R_{i,j} \right),$$

where N is the number of the possible states in the phase space and graphically corresponds to the number of the pixel in our picture.

The RR for all the stimulation are shown in Table 9.4 and it reveals that the homogeneity between the subjects is always under the 23 % with threshold $\varepsilon = 0.3$ and it reach the 36.5 % with threshold $\varepsilon = 0.5$.

Fig. 9.16 How RP works

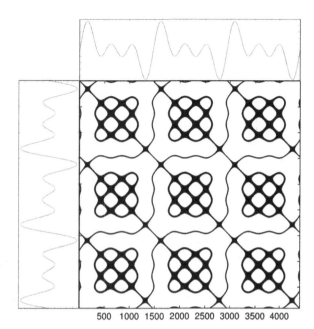

9.3.3   Explicit Approach: Conscious Production of Single Notes

With the pure aim to test if we could make users able to produce sounds by their brain using a BCI, we realized a prototype version of an entertainment tool allowed users to create music through brainwaves (Folgieri and Zicchella in press). Other researchers (Dan et al. 2009; Hamadicharef et al. 2010) are exploring this field, creating application for music composition. Our main objective was to evaluate how reducing as much as possible the training time needed by BCI devices.

Materials and Methods

For the realization of the music application prototype (Folgieri and Zicchella in press) we used Processing[2], an open-source programming environment allowing the development of a Java application realizing the connection between Processing, the chosen BCIs, Emotiv and Mindwave, SDKs and Max 6[3], a popular environment for visual programming, specifically developed by Cycing '74 for applications in music and multimedia. We added a few lines of code to process the collected BCIs' signals and to manage alpha, beta, gamma, delta brainwaves and eye blink. In the exploratory experiments we excluded brain activity related to eye blink, but, when

---

[2] http://www.processsing.org.

[3] http://cycling74.com/products/max/.

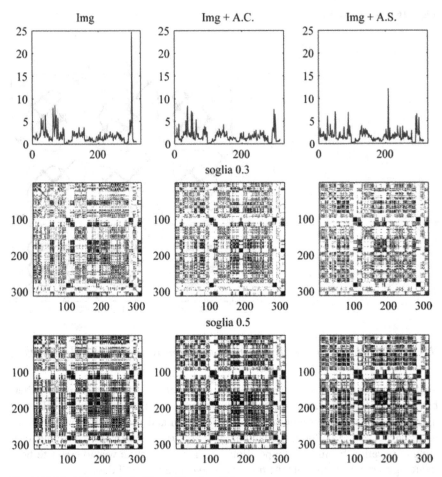

Fig. 9.17 The RP computed with ε=0.3 and ε=0.5

applying our findings in developing an entertainment prototype application, we included also this signal, either considering that we aim to use commercial devices in a real-world environment, and on the basis of eye blink and emotion studies, such as, for example, (Jackson et al. 2003). The stimuli used to reduce users' training consisted in single music notes, reinforced by symbols.

Preliminary Experiment Organization

In a first phase, we performed a preliminary experiment choosing seven subjects, three women and four men, aged between 14 and 49. Difference in age has been considered potentially relevant for the variability of the results. Once the prototype was developed, we successfully tested it on 50 individuals, aged between

	Set	Img	Img + A.C.	Img + A.S.
$\varepsilon=0.3$	1	21.3819	20.6977	17.8357
	2	18.3987	19.0233	19.0417
	3	20.8600	21.6120	20.0999
	4	19.3438	18.1377	17.6775
	5	18.4377	19.4321	22.6866
	6	19.5924	21.1580	18.4788
	7	18.2220	18.8651	20.8107
$\varepsilon=0.5$	1	34.4305	33.4874	29.1132
	2	29.8714	31.1904	31.0651
	3	33.8942	34.7798	32.4499
	4	31.4472	29.1831	28.5832
	5	30.0563	31.7041	36.5631
	6	31.7205	34.1757	30.1446
	7	29.9659	31.5288	33.8798

Table 9.4 Recurrence Rate of the responses of the subjects for each stimulation set for two different values of threshold ε

14 and 52, equally distributed in male and female, musicians and not. Each participant took the test separately, in a comfortable environment, to reduce variation influenced by external diseases, wearing the BCI and headsets, completely isolated from the external world remaining concentrate on the sound listened in the headsets (the single note). Each EEG registration session had the duration of two minutes, during which subjects were resting, recommended to not close their eyes, do not speak or move.

The experiment consisted in verifying if subjects could be trained to reproduce a single specific note, through a BCI device and the developed interpretation software. Participants have been invited to reproduce more single notes, supported by a presented audio stimulus.

The first attempt to make subjects able to reproduce the listened notes revealed that the needed training time was too much. The difficulty, in fact, was not by the computer side in interpreting the subjects' brain signal, but on the subjects' side for the difficulties to focus only on the sound, without distracting.

Other attempts to reach the success in making subjects able to reproduce with their mind the target sound (the single note) demonstrated that the subjects needed to try more than 3–4 times, before having success. Then we thought to apply a reinforce stimulus to help subjects.

The solution consisted in requesting to participants to make a gesture while observing an image and listening to the note for one second, choosing for each note a specific gesture and a specific image. Adopting this solution, the training time has been reduced, in all the cases, of at least 40–50 %. The following step, to refine the developed tool, has been to set up a sketch in Processing to use the discovered characteristics.

Once the software with the individuated refinements was prepared, the participants were invited to think of the observed colors and of the note listened to previously, associating the corresponding motor stimuli, following the instructions.

Results

Main result consists in observing that a great part of the variations in EEG signals occur in beta (associated with active thinking, active attention, focus on the outside world or solving concrete problems) and alpha (related to relaxed awareness, meditation, contemplation) bands.

We obtained that, after a relatively short training and with the help of a visual and a motor stimulus, the subjects have been able to reproduce the requested notes. This technique is often used in a similar way to train the users in executing virtual actions on a computer, such as, for example, the rotation of a cube or, linking a BCI to an electronic device, to control it.

We obtained a prototype version of an entertainment tool allowing users to create music through brainwaves (more details inside (Folgieri and Zicchella in press)). The application implements the following characteristics: (a) each note (to be reproduced after) is listened just one time by the users; (b) before the reproduction of the target note, the program gives to the user the instruction asking to associate a simple gesture to the listening (the software suggests a different gesture for each note); (c) at each listening of a note the software shows the associated image (the name of the note on a different note-specific color background). Thanks to this system a generic user can be trained, listening all the seven notes in the same session, and, after, correctly reproduce them just evoking the associated sounds, images and gestures. For the most part, users are able to reproduce from three to all notes at the first use of the software.

Apart from obtaining the first prototype for creating sound by brainwaves, the experiment demonstrates a correlation between the execution of an action and the will to execute it, also when the action is mainly non-motor.

9.3.4   *Work-in-Progress and Furhter Developments*

As told, one of our research objectives is to test the possibility to influence games' music following or contrasting the players' mental state to enhance their emotional experience and involve them more in the game. With the aim of exploring this possibility in enhancing users' fun, we have created a VR environment reproducing scenes from famous games (Call of Duty, Resident Evil and Guitar Hero) to track the four mental states detected in our preliminary study phase (i.e. frustration, satisfaction and level of arousal due to stress or relaxation) from users wearing the Emotiv Epoc or the Neurosky MindWave. Our aim consists in testing the possibility to regulate games sound and music following user emotion changing. Depending on the passively detected user's mental state, we developed an algorithm operating on the intensity of the sounds (increasing with low arousal and high frustration, decreasing on the contrary); changes of background music (for example contrasting, with fast/low rhythms and music genre, the stress and/or the satisfaction level of the user). Currently, the experiment is in progress, and we planned to test our algorithm on about 60 individuals, mostly students of our University Department. Preliminary results showed promising progresses in our research. The final aim consists in creating a standard module, easy to interface to games and eligible for games farms.

We are also planning to investigate other applications deriving from the obtained results. Recent advancements in BCIs allow one to interface them to mobile devices, so we are currently working also on a "neuro" version of a playlist and shuffle mechanisms. We are investigating the possibility of providing a user for a random compilation selected among music tracks labeled on the basis of EEG signal (or other biofeedback signals detected by a BCI) registered during the listening of the tracks and related to specific emotional state (emo-tagging). The user could choose to follow or contrast own current emotional state.

9.4 Conclusions

In preliminary experiments performed to test our approach, results showed that brain activity measured at the anterior part of the scalp distinguished the valence of musical emotions both using the Emotiv Epoc and the Neurosky MindWave. In addition, the activity increases during the presentation of positive-valence musical excerpts, while it decreases correspondence to negative ones. The overall frontal activation is related to the intensity of emotions elicited by music: it decreases corresponding to sound related to fear and frustration and increase for happiness or satisfaction.

We also found that, both the Emotiv Epoc and the Neurosky MindWave have the sufficient number of sensors needed for our objectives. Therefore, the Emotiv Epoc showed a greater precision in detecting features individuating the selected users' mental states (89 vs. 81 %), probably due to the direct measure of the hemisphere. Moreover, the Neurosky MindWave is preferred by users for its wearability.

Considering the results and the obvious correspondence between music and emotions, the adaptation of the game's sounds to the emotional and mental state of a player detected by a BCI appears a promising application. In fact, BCIs can be easily used to detect if the user is bored or inattentive, consequently modifying the music track or sounds to involve again the player in the game scenario. Concerning exclusively music entertainment, consider, for example, the famous game "Guitar Hero": thanks to a BCI device it could be possible to modify the music genre, the rhythm or the intensity of the sounds following the variability of the user's emotional states, introducing a new modality in game level achievement.

We also developed a prototype of an application allowing users to play single music notes by their brainwaves. The main aim of our experiment consisted in this case in developing a tool for conscious production of music by the brain reducing, through appropriate stimuli, the training time needed by individuals and allowing to a generic user to reproduce any single note. The results of the experiments gave us the possibility to verify that with the alone EEG signal, a subject could need a long training. In fact, only every 3–4 attempts were subjects able to correctly reproduce the note. Better results have been obtained if the listening of the note we associated a motor and a visual stimulus. In such a way, in fact, we potentiate the subjects' ability to concentrate on the task, with a consequent increasing and differentiation in beta waves. Consequently, a subject is able to reproduce the specific target note. The developed software included these results, so the needed training time has been strongly reduced.

References

Allison BZ, Wolpaw EW, Wolpaw JR (2007) Brain-computer interface systems: progress and prospects. Expert Rev Med Devices 4(4):463–474

Ansari-Asl K, Chanel G, Pun T (2007) A channel selection method for EEG classification in emotion assessment based on synchronization likelihood. In: Eusipco 2007, 15th European signal processing conference, Poznan, Poland, 2007

bin Yunus J et al (2012) The effect of noise removing on emotional classification. In: IEEE Computer & Information Science (ICCIS), 2012 International Conference on, vol 1, pp 485–489

Brown JC (1991) Calculation of a constant q spectral transform. J Acoust Soc Am 89:425

Call of Duty. Retrieved 23rd February, 2014, from http://www.callofduty.com

Choi S, Cichocki A, Park HM, Lee SY (2005). Blind source separation and independent component analysis: A review. Neural Information Processing-Letters and Reviews, 6(1), 1–57.

Comon P (1994) Independent component analysis, a new concept? Signal Process 36(3):287–314

Dan W, Chao-Yi L, De-Zhong Y (2009) Scale-free music of the brain. PloS ONE

Davidson RJ, Schwartz GE, Saron C, Bennett J, Goleman DJ (1979) Frontal versus parietal EEG asymmetry during positive and negative affect [Abstract]. Psychophysiology 16:202–203

Szibbo D, Luo A, Sullivan TJ (2012) Removal of blink artifacts in single channel EEG. In Engineering in Medicine and Biology Society (EMBC), 2012 Annual International Conference of the IEEE:3511-3514. IEEE.

Eckmann J-P, Olifson Kamphorst S, Ruelle D (1987) Recurrence plots of dynamical systems. Europhys Lett 4(9):973–977

Emotive System Inc. Retrieved 23rd February, 2014, from http://www.emotiv.com

Folgieri, R, Zichella M (2012) Conscious and unconscious music from the brain: design and development of a tool translating brainwaves into music using a BCI device. In: Proceedings of the 4th International conference on applied human factors and Ergonomics, San Francisco, USA, CRC press.

Gaetano Valenza AL, Scilingo EP (2012) The role of nonlinear dynamics in affective valence and arousal recognition. IEEE Trans Affect Comput 3(2):237–249

Garcés Correa A, Laciar E, Patiño HD, Valentinuzzi ME (2007) Artifact removal from EEG signals using adaptive filters in cascade. J Phys: Conf Ser 90:012081 (IOP Publishing)

Hamadicharef B et al (2010) Brain-Computer Interface (BCI) based musical composition. In: CW2010, pp 282–286

Hamilton JD (1994) Time series analysis, vol. 2. Princeton University Press, Princeton

Huang D, Zhang H, Ang K, Guan C, Pan Y, Wang C, Yu J (2012, March). Fast emotion detection from EEG using asymmetric spatial filtering. In Acoustics, Speech and Signal Processing (ICASSP), 2012 IEEE International Conference on, IEEE, pp. 589–592

Hyvärinen A, Oja E (2000) Independent component analysis: algorithms and applications. Neural networks 13(4):411–430

Jackson DC, Mueller MJ, Dolski I, Dalton KM, Nitschke JB, Urry HL, Rosenkranz MA, Ryff CD, Singer BH, Davidson RJ (2003) Now you feel it, now you don't: frontal brain electrical asymmetry and individual differences in emotion regulation. Psychol Sci 14:612–617

Keil A, Müller M, Gruber T, Wienbruch C, Stolarova M, Elbert T (2001) Effects of emotional arousal in the cerebral hemispheres: a study of oscillatory brain activity and event-related potentials. Clin Neurophysiol 112(11):2057–2068

Kirmizialsan E, Bayraktaroglu Z, Gurvit H, Keskin Y, Emre M, Demiralp T (2006) Comparative analysis of event-related potentials during Go/NoGo and CPT: decomposition of electrophysiological markers of response inhibition and sustained attention. Brain Res 1104(1):114–128. doi:10.1016/j.brainres.2006.03.010

Ko M, Bae K, Oh G, Ryu T (2009) A study on new gameplay based on brain-computer interface. In: Barry A, Helen K, Tanya K (eds.) Breaking new ground: innovation in games, play, practice and theory. Proceedings of the 2009 digital games research association conference, Brunel University

Lalor EC, Kelly SP, Finucane C, Burke R, Smith R, Reilly RB, McDarby G (2005) Steady-state VEP-based brain-computer interface control in an immersive 3D gaming environment. EURASIP journal on applied signal processing, 2005, 3156–3164

Laurier C, Sordo M, Serra J, Herrera P (2009) Music mood representations from social tags. In: Proceedings of international society for music information retrieval conference (ISMIR), Kobe, Japan

Lee TW (1998). Independent component analysis (pp. 27–66). Springer US.

Lu D, Liu L, Zhang H (January 2006) Automatic mood detection and tracking of music audio signals. In IEEE transactions on audio, speech and language processing, vol 14, no 1, pp 5–18

Metal Gear Solid. Retrieved 23rd February, 2014, from http://www.konami.jp/kojima_pro/english/index.html

Morris JD (1995) SAM: the self-assessment Manikin, an efficient cross-cultural measurement of emotional response. J Advertising Res

Mühl C, Gürkök H, Plass-Oude Bos D, Thurlings ME, Scherffig L, Duvinage M, Elbakyan AA, Kang S, Poel M, Heylen DKJ (2010) Bacteria hunt: a multimodal, multiparadigm BCI game. In: Proceedings of the international summer workshop on multimodal interfaces, Genua

Müller M, Keil A, Gruber T, Elbert T (1999) Processing of affective pictures modulates right-hemispheric gamma band EEG activity. Clin Neurophysiol 110(11):1913–1920

Murugappan M, Nagarajan R, Yaacob S (2009) Appraising human emotions using time frequency analysis based EEG alpha band features. pp 70–75

NeuroSky Inc. Retrieved 23rd February, 2014, from http://neurosky.biz

Petrantonakis PC, Hadjileontiadis LJ (2012) Adaptive emotional information retrieval from EEG signals in the time-frequency domain. IEEE Trans Signal Process 60(5):2604–2616

Pineda JA, Silverman DS, Vankov A, Hestenes J (2003) Learning to control brain rhythms: making a brain-computer interface possible. Neural Systems and Rehabilitation Engineering, IEEE Transactions on, 11(2), 181–184

Plass-Oude Bos D, Reuderink B, Laar B, Gürkök H, Mühl C, Poel M, Nijholt A, Heylen D (2010) Brain-computer interfacing and games. In: Tan DS, Nijholt A (eds) Brain-computer interfaces, series human-computer interaction series

Resident Evil. Retrieved 23rd February, 2014, from http://www.capcom.co.jp/bio5/

Russell JA (1980) A circumplex model of affect. J Pers Soc Psychol 39(6):1161–1178

Ryu CS, Song Y, Kim SH, Yi I, Kim JE, Sohn JH (1998). A time-frequency analysis of the EEG evoked by negative and positive visual stimuli. In Engineering in Medicine and Biology Society, 1998. Proceedings of the 20th Annual International Conference of the IEEE, IEEE, Vol. 4, pp. 2012–2015

Schmidt LA, Fox NA (1999) Conceptual, biological, and behavioral distinctions among different types of shy children. In: Schmidt LA, Schulkin J (eds) Extreme fear, shyness and social phobia: origins, biological mechanisms, and clinical outcomes. Oxford University Press, New York, pp 47–66

Schmidt LA, Trainor LJ (2001) Frontal brain electrical activity (EEG) distinguishes valence and intensity of musical emotions. Cognition Emotion 15:487–500

Scott M, Anthony JB, Tzyy-Ping J, Terrence JS et al (1996) Independent component analysis of electroencephalographic data. Adv Neur Inform Process Syst :145–151

Silent Hill. Retrieved 23rd February, 2014, from http://www.konami.com/games/shh/

Sniper Elite. Retrieved 23rd February, 2014, from http://www.microids.com/en/catalogue/28/sniper-elite-berlin-1945.html

Stone JV (2005) Independent Component Analysis. Encyclopedia of Statistics in Behavioral Science. Wiley Online Library

Tomarken AJ, Davidson RJ, Wheeler RE, Doss RC (1992) Individual differences in anterior brain asymmetry and fundamental dimensions of emotions. J Pers Soc Psychol 62:676–678

Tseng KC, Wang YT, Lin BS, Hsieh PH (2012, July). Brain Computer Interface-based Multimedia Controller. In Intelligent Information Hiding and Multimedia Signal Processing (IIH-MSP), 2012 Eighth International Conference on, IEEE, pp. 277–280

Zhou W, Gotman J (2009) Automatic removal of eye movement artifacts from the eeg using ica and the dipole model. Prog Nat Sci 19(9):1165–1170

Chapter 10
Computer and Music Pedagogy

Kai Ton Chau

10.1 The Quest

A number of years ago, a discussion between several music educators, who were also computer enthusiasts, fascinated me. The discussion was about the use of computers to enhance learning in basic musicianship, particularly pitch and rhythm matching—some of the fundamental music abilities for singers.

> When a student is asked to match a tone (pitch) generated by a computer, how does the computer determine if the response is accurate? In the same way, when a student is asked to repeat a rhythmic pattern generated by a computer, how would the computer determine the accuracy of the response?

To the unsuspecting, the processes of the above events might seem quite straightforward: In the former case, the computer is programmed to generate a tone, and seeks a response from the user. When the student responds by repeating the pitch, the computer will capture the sound from an input source (typically a microphone), and then analyze it (particularly, the frequency of the tone recorded). If the numerical value of the frequency captured by the computer matches with that of the preset value of the underlying question, the answer would be correct, and thus the computer would provide a positive feedback to the student. If not, the feedback would be a negative one. In the latter case, the computer generates a series of rhythmic patterns within a defined speed (tempo) chosen by the programmer. The response from the student is again captured by a microphone, and then analyzed by the computer. If the length of each note is an exact match with the preset patterns, the answer would be correct; if not, the response is not an accurate one.

Represented logically, if a represents the numeric value of the frequency or rhythmic pattern generated by the computer, and b represents the numeric value of the frequency or rhythmic pattern captured by the computer from the response, then:

K. T. Chau (✉)
Kuyper College, Michigan, USA
e-mail: kaiton.chau@gmail.com

N. Lee (ed.), *Digital Da Vinci*, DOI 10.1007/978-1-4939-0536-2_10,
© Springer Science+Business Media New York 2014

when a = b; the response is accurate, and
when a ≠ b; the response is inaccurate.

It all sounds great; at least in theory.

In reality, however, the human voice consists of very complex sound patterns. These patterns are far from the purity of sine waves. Only with a great deal of training and intentional effort, an experienced singer can produce a "pure" vowel tone with a more-or-less consistent wave pattern and frequency. Even if the wave patterns are stable, what about the frequency of the pitch of the response? Let us ask ourselves this question: If a computer were to be programmed to generate the pitch of an A4 (at a frequency of 440 Hz), would a human response of exactly 440 Hz be the only acceptable "accurate" response? Granted, well trained human ears can detect the difference of 1 Hz; it takes a trained singer to produce a "straight tone" of a particular pitch for a sustained period of time. (A "straight tone" is a musical tone without vibrato. Vibrato is an intentional technique which allows the tone be produced in a small range of frequencies above and below the intended pitch. Musically speaking, a well-used vibrato adds warmth and texture to the intended note. In vocal music, the technique of vibrato helps relax the larynx while singing. In contrast, a singer with a "straight tone" sustains on one pitch, that is, the intended frequency.) For the computer to determine if the response is a match, the task boils down to two main criteria. One, since the human voice may not be producing a pitch that is exactly at 440 Hz, how much "off-pitch" should the computer consider the response "accurate"? In other words, how much *tolerance* in terms of plus-and-minus a number of Hz or a percentage variation should the computer accept as an *accurate* response? Two, if the response is not a "straight tone," which segment(s) of the response should the computer accept and perform its evaluation. Sound functions in the concept of time: If there are variations of frequencies in the response, which segment (in terms of time) should the computer choose? While any of the segments, no matter how long, may not be representative of the full length of the response, an average ("straightening out") of the frequencies does not make much more sense either.

The above conversation, in fact, occurred in the mid-1980s, more than 30 years ago at a time when the microcomputers were gaining ground in the business world. With the advent of graphic computing on the microcomputer platform around that time, and the development of hardware (such as the sound card) needed for multimedia applications, educators started to wonder how computer could enhance learning. Music educators have been experimenting in many areas of employing the computer and technology in music pedagogy ever since.

10.2   Technology and the Music Classroom

The music classroom has experienced significant change since the mid-1960s in terms of the employment of technology. Peter R. Webster of Northwestern University defines *music technology* as "inventions that help humans produce, enhance, and better understand the art of sound organized to express feelings (Webster 2002)." According to his definition, the goal of music technology is three-fold: to produce music, to enhance the expression of music, and to better understand music. We have witnessed how technology has changed the production of music. In just a few decades, we have moved from the ability to capture sound in an analog or electromagnetic way on vinyl records and magnetic tapes, to digital media. At the same time, the notation systems have made such advancements that have made pens and pencils and engraving obsolete very quickly. Technological changes in music production have brought about many opportunities to the music education arena. For example, many music schools now offer music composition software for the purposes of doing simple assignments to full-scale orchestration, with some universities even equipping their facilities with industry-level recording studios.

In terms of the expression of music, technology has opened a wide spectrum of sounds that were unimaginable just a generation ago. A simple electronic keyboard from several decades ago, for example, was capable of making a full range of sounds that imitated quite well with that of a piano, an organ, or a harpsichord. With computerized sampling capability developed in the 1970s, a keyboard could not only generate musical tones that are virtually undistinguishable from hundreds of musical instruments, but also produce sounds of clapping, bird chirps, a locomotive, a kiss, and even simulating a singing human voice! Many electronic keyboards are capable of producing the full sound of a band—with drums, harmony based on all common chords (major, minor, diminished, all varieties of seventh chords, and more), and a number of solo instruments. For music educators, they may now instantly demonstrate many musical effects for the students; and for the students, they may get instant feedback on the music ideas they create. As prompt feedback is a crucial part of learning, technology has brought much enhancement in the cycle of acquiring the knowledge—praxis—feedback—improvement. Having studied music in college during the early 1980s myself, I witnessed the journey of learning orchestration, which involved listening to many different passages of different combinations of instruments (for example, how a flute and an oboe, or the French horn and bassoon, would sound when they pair up and play together), learning the ranges and characteristics of all kinds of orchestral instruments, and then arranging a piano sonata for a full orchestra. A sizable amount of imagination was needed (in addition to many rules) as I assigned certain passages to an instrument or a group of instruments. My teacher played a key role in helping us do things the "right" way. The fact was, it was not practical logistically and financially to have an orchestra at our disposal to play our work-in-progress, nor could we afford an orchestra to play ineffective assignments of college students. As we approached the end of the semester, the big day came when we had our arrangement ready to be rehearsed by a full orchestra after countless long nights of copying part scores by hand, using nothing but pencil and paper. (Thanks to the invention of the photocopier, we didn't have to prepare multiple copies for the string section by pencil and paper.) It was an overwhelming experience—for lack of a better expression—to hear our own work played by a full orchestra. For those who had journeyed through this process, you may share my own experience that even if you wanted to make some changes to the orchestration based on the actual sound you heard at the first rehearsal, there weren't a lot of things you could modify at that point. Fast forward a decade or so, and you can see many students being able to afford to have their own electronic keyboard that is attached to a computer. Not only are they able to write and modify notation very easily, the MIDI technology can afford them to listen to a simulated effect that is very close to the actual symphonic orchestra in a concert hall setting. Not to mention that once the full score is ready, the computer can generate part scores with the touch of a button. Technology has truly changed the landscape of the production and expression of music.

Technology enhances the understanding of music through new pedagogy. Pedagogy is both the art and the science of teaching. Music, as an expression of emotion, is as ancient as human history. As one person passes on his finding and understanding to another, the process involves teaching. Pedagogy, obviously, involves a more intentional methodology in teaching and passing on information. Modern research on pedagogy show that learning can be enhanced by employing a variety of teaching methods; some examples of teaching methods include repetition, using multiple ways or media to explain a concept, addressing the needs of a student's learning styles, etc. The versatility and affordability of modern computers have opened many new possibilities.

Since the 1960s, scholars have done extensive research on music education technology, with one of the more notable articles from this era being "Computer-Assisted Teaching: A New Approach to Research in Music" by Wolfgang E. Kuhn and Raynold L. Allvin (Kuhn and Allvin 1967). In the mid-1990s, William L. Berz and Judith Bowman co-authored a book entitled *Applications of Research in Music Technology*, in which they studied the use of computer-aided instruction (CAI) in music education since the publication of Allvin's work in 1967 (Berz and Bowman 1994). In a blog published by Alex Ruthmann of the New York University, he regards Berz and Bowman's book as "influential to educators" because "it offered suggestions for applying the results of the research to practice." (Ruthmann 2006) Webster, as mentioned earlier, extended the research timeframe to 2002, and then again 2005. In his 2005 study, Webster reviewed more than 60 research projects on computer-based technology and music teaching and learning between 2000 and 2005, and he raised the concern of the need of "more substantial studies on teaching strategies that use technology, among other issues (Webster, Computer-Based Technology and Music Teaching and Learning 2000–2005 2007)." Computer technology has since become a forefront discussion in music pedagogy.

Today, there are many computer applications capable of teaching music. Some applications focus on drilling on music rudiments, while others provides a platform for self-practice. Many more educators use general computer technology (such as telecommunication capability involving video and sound, or various social media platforms) to enhance teaching and learning of music. For the rest of the article, I attempt to summarize the various methodologies into four major categories—the right-or-wrong approach, applications for practice, visualization of music concepts, and general technology use. I will share some of my and my colleagues' classroom experiences in some of these areas. At the end, I will also try to look at the technological trends of music teaching in the near future.

10.3   Approach in Music Learning

10.3.1   The Right-Or-Wrong Approach

A more traditional way of using computers for teaching music is question-and-answer drills. In essence, the designer and programmer of the application would prepare a large quantity of questions together with their corresponding answers. The computer picks a question randomly or according to a progressive scheme of difficulty levels, and then waits for a response from the user. The user response could take a variety of formats—fill in the blanks with specific key strokes, multiple choice, and true-or-false are some of the examples. The computer is preprogrammed to expect a certain definitive value of response from a user. The computer then compares the response with the preprogrammed value—a match of these two values would indicate a correct response, while a mismatch would be an incorrect response. This approach of teaching, in a sense, is similar to arithmetic exercises that focus on the likes of addition or subtraction, or even spelling exercises. The user must provide a definitive input. The computer evaluates the answer by comparing the input with the pre-determined answer or formula set up by the programmer. The multiple-choice approach expands the horizon of the questions by providing four or five descriptive or qualitative sentences, and allowing the user to choose a specific option. The multiple-choice format has been used for many years and by many disciplines—from academic subjects to driver licensing. In fact, a number of online learning systems have the functionality built in so that the course instructor can design his/her own multiple-choice tests easily.

No matter what the input format is, the strategies of these computer-based exercises are based on the evaluation of the user input to be right or wrong, and repeated drilling.

Music, in its very nature, is a distinctive language shared by the professionals and amateurs alike. In musical notation, it uses special symbols that have specific meanings attributed to them. Some of the examples include the grand stave, the various kinds of clefs (treble clef, bass clef, alto clef, tenor clef, etc.), and a system of duration and rhythmic values of notes. Because of the symbols involved in the musical language, the development of graphic displays for microcomputers in the 1980s and 1990s opened up the possibilities of having drill questions based on the graphic nature of the musical language. The graphic capability of microcomputers—first the color graphics adapter (CGA) introduced in 1981, and then the much more capable video graphics array (VGA) in 1987—could display a key signature such as 𝄞♯♯♯♯ and expects an answer of C ♯ major or A ♯ minor. Because of the popularity of the use, certain musical symbols in fact have their own Unicode or HTML codes. The followings are some common examples:

Musical symbol	Unicode	HTML code
♯ (sharp)	U+266F	&#9839
♭ (flat)	U+266D	&#9837
♮ (natural)	U+266E	&#9838

This right-or-wrong approach can be used for many kinds of questions. The basic questions on musical rudiments may include the recognition of notes and letter names, key signatures, scales, intervals, and chords. In my music rudiments class, I have used the web browser based musictheory.net (http://www.musictheory.net/exercise) and an Android app called *eartrainer3* published by Clemens-Alexander Brust IT Dienstleistungen, Germany. The user interface (UI) of these applications is simple and intuitive; students of any age can participate in the learning very quickly.

Like the music labs in many higher learning institutions, the music lab in our college is equipped with computers and electronic keyboards. Each keyboard is connected to a computer as an input source. With a connected keyboard, a student may practice on finding the right notes on the keyboard that correspond to the displayed notes on the computer monitor. This kind of combined visual, aural and kinesthetic learning is particularly helpful for beginning learners of all age groups on music rudiments.

From a pedagogical viewpoint, the use of computer-based question-and-answer drills has several advantages. First, the computer application provides instant feedback. Timely feedback is one of the key elements for effective teaching. The instant feedback helps students learn while the question is still fresh in their mind. When the students make mistakes, the instant feedback from the computer application allows the students to rapidly review their mistakes, and learn from them quickly. In contrast, a paper-based quiz would take the instructor a day or two to review, and has to wait for the next class time to be sent back. The instant feedback shortens the turn-around time, and therefore let the students move on with their learning more efficiently. Second, the computer-based exercises and quizzes can extend the learning experience to the student's home. Using the appropriate computer applications, the learning can continue to be interactive even when the student is not in a classroom setting. Third, well-designed computer applications usually stratify learning based on difficulty levels. When the students perform satisfactorily at a certain level, they advance to a higher difficulty level. The approach can keep the students engaged and challenged, while give them a sense of achievement when specific difficulty levels are completed. The stratification approach may also be used to cater the individual needs of the students—beginners may do more exercises at the foundational levels, while the learners with more understanding of the subject can start at a more advanced starting point. Other benefits of computer-based Q&A drills also include the likes of randomization and repetition of questions, as well as a more effective use of the instructor's time in teaching administration.

The teaching of music, particularly in music theory, has come a long way since the 1960s when electronic technology in music had become more accessible and affordable. In 1992, W. Higgins reported the expanse technological and computer usage in music teaching and learning, which included the use of television, computer, and electronic keyboard with MIDI technology (Higgins 2002). The technological advancement has since revolutionized the teaching of music in the classroom.

10.3.2 Practice, Practice, and Practice

Another type of computer usage in music teaching and learning is practice aide. In this respect, the computer technology can play a wide variety of roles—from providing a click track to a full accompaniment.

Conventional wisdom says, "Practice makes perfection." (Someone may argue that practice does not necessarily lead to perfection—it only leads to permanence! In this sense, to practice *correctly* is as important as to practice often. At the present time, computer technology may not be able to improve one's performance style, but it may assist the practitioner in terms of steady beat or practicing in a different tempo.) At a very basic level, a metronome helps musicians to keep their beat. An electronic metronome has several advantages over a traditional mechanical metronome such as portability and preciseness. Portability refers to the ease of carrying the device around, from practice studio to a concert venue. Electronic metronomes have better preciseness over the mechanical counterpart because they do not need a horizontal surface to function properly, and the tempo increments are more defined. Computer-based portable metronome, particularly those programmed as iPad or Android apps, can make tempo increments as low as one beat per minute. Some apps are capable of doing a reverse metronome calculation, which means that when a user provides a steady tap pattern on the touch screen, the app then provides a precise metronome count for the pulse. Meanwhile, some of these apps often come with tuning utilities that are not limited to A=440 Hz, but the full chromatic scale.

Software-based click tracks could be either prerecorded pulses (or clicks) or programmable metronome beats. These tools are widely used in band practice, live performances, and in the film industry. Computerized click tracks help bring steady tempo to a group of performers, whether experienced or not. Some of these systems are very elaborate that are capable to include multiple channels, the sampling of existing wave (.wav) files, and presetting sequence of songs. Some simply provide a steady pulse, controlled by a member of the band (usually the drummer), such that every member of the band stay in the same tempo.

In a different scenario, computers have opened up new possibilities for musicians to practice their instruments. Music professionals and students alike spend many hours in practicing their instruments, voice, or conducting. Certain genres of music, such as concertos, quartets, arias, etc., call for collaboration among several or even a large group of musicians. For many music students, having an orchestra available for them to practice a concerto would be ideal, except that the arrangement may not be financially practical. To engage a full symphonic orchestra (with 40 to 60 musicians or even more) is a financial burden for most students, the reasons being (1) the number of orchestral musicians and the payments involved, and (2) that the orchestral musicians themselves need practice before collaborating with the solo musician, thereby adding further expenditures to the arrangement. In a practical sense, most student performers practice with a pianist who plays from a reduction score (that is, a piano reduction of the full orchestra score). This traditional practice method has helped many student performers practice their performance pieces in an aural partnership with the accompanist, and also enjoy the flexibility of working with a live accompanist.

Band or orchestra students in the middle or high schools also have a need to practice as a group. While it is important for the full band and orchestra to come together as a full group to practice together, educators agree that individual practice before the full band or orchestra session is crucial to the success of the group learning experiences. Traditionally, the music student would go home or to a practice room and practice their individual section. Although the individual practice is invaluable in the learning process, the student lacks the interactive elements of working with the rest of the group—such as the preciseness of the tempo setting and the dynamic changes.

The advancement of technology has opened up a new pedagogical approach to music practice. A few decades ago, Ira Krata pioneered the idea of "Music Minus One" (or abbreviated as "MMO") (www.musicminusone.com)—a recording of a piece of music while the main instrument is excluded. MMO is used as an aide to practice. The very same idea is used in *karaoke* for the voice. For example, an MMO production could feature a recording of a piano concerto with the solo piano part excluded. Or, in the case of a quintet, all five instruments are omitted in five different MMOs. (This was the case for Franz Schubert's Trout Quintet—the first recording released by Music Minus One in 1950.) The idea received critical coverage in the media, including Life, Time, Newsweek, and many other publications across the United States and Europe. MMO has since proven to be an effective pedagogical tool for students and amateur musicians.

While MMO is helpful for practice, it lacks flexibility in terms of tempo and key. These factors are obvious because it is impossible to manipulate the speed (tempo) of a performance prerecorded on a cassette tape, a vinyl record, or a CD, without sacrifice to the integrity of the performance itself. If a student wants to play the practice piece slightly slower, the lower play speed on the tape player or the record player can change the pitch (or the key), and the intonation of the musical piece! The computer age has brought the concept another big step forward. What if the students may choose the tempo so that they may start with a more manageable tempo and then gradually move up to the expected tempo? What if the students may isolate the more difficult passages and practice repeatedly? What if they can get an instant feedback on the right and wrong notes they played, and a recording for their practice? What if the teacher is able to listen to their practice and give the students feedback (and grades)? These what-ifs are now possible with a newer generation of online practice software, one of them being SmartMusic (www.smartmusic.com). This popular educational software is often used in schools, particularly in bands and orchestras. *Interactivity* is the key strength of such kind of software; students can not only have a "full accompaniment" of the band or orchestra, but also control the tempo and whether to use a click track to aid their practice. This kind of software is an excellent pedagogical tool because the teacher may assign segments of music for students for their daily practice. The students may submit a recording (or their *best* recording) to the teacher over the Internet as a proof of their daily practice. Over time, the teacher may compile these recordings to showcase the progress of the students—an encouragement to the students indeed!

Back in the 1980s I was a college student in music composition and later, a graduate student in choral conducting. Like instrumental players, one of the challenges in practicing conducting was having a group of students or volunteers willing to spend time in our journey of learning. Playing prerecorded performances on records or CDs while we conducted was a good practice method for gaining confidence in technique and style, but records and CD fell short on flexibility as we attempted to infuse some personal preferences and interpretations such as tempo and dynamics changes. In the mid-1980s, our conducting classroom was the piano lab. While one student practiced conducting, the rest of the class became the "orchestra" as we played out the parts on the electronic pianos. Nowadays, realistic orchestral sound may be achieved with MIDI and "sound libraries" software. With capable computer software and a competent keyboard player, practice has become a lot more interactive, responsive, and flexible.

There is a saying that goes like this: Imagine if you are lost in New York City on your way to Carnegie Hall, and you see someone who looks like a musician. You ask the musician for direction, "How do you get to Carnegie Hall?" The answer you get would probably be, "practice, practice, and practice." Serious musicians spend a lifetime in practice in order to perfect their skills, interpretation, and collaboration. Computers as a pedagogical tool could make practice more interactive, effective, and fun!

10.3.3 Sound Visualized (Demonstration)

I teach a course titled "Introduction to Music" to non-music majors as an elective. In additional to the appreciation of music from several style periods, genres, and from several world cultures, the course also emphasizes critical thinking and the ability to express why one likes or dislikes a certain kind of music. To enable a meaningful discussion, the first learning objective of the course is to obtain a basic understanding of the musical language, which covers the four fundamental elements of a musical tone—pitch, duration, loudness, and timbre. The exploration then moves on to the formation of a melody, harmony, and later on the form and the structure of music.

To many trained musicians or music lovers, the elements of the musical language come like second nature—we sing and play a wide range of pitches and loudness, and we intuitively make musical sound of different duration and timbre. But sometimes it is difficult to explain the musical language using musical language itself— that is very much like trying to explain a foreign language (which is *foreign* to the party who you are talking to) with the very same foreign language. In fact, it could be a tough task to explain the difference of a high pitch and low pitch to a musically challenged, otherwise known as "tone-deaf," student. An oscilloscope would be a helpful instrument for both the visual learners and the kinesthetic learners. For the former, the oscilloscope displays visually the frequency of the sound wave; for the latter, they can make a sound to the microphone and learn with hands-on experiments. Due to tight budgets experienced by many schools and colleges, an oscilloscope may be out of reach for the music appreciation course. Thanks to computer software and tablet apps (some of these applications are even free), a simulated

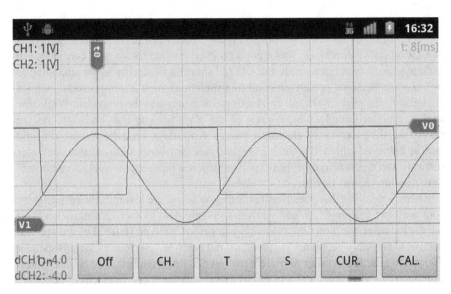

Fig. 10.1 A screenshot of the OsciPrime Android app

oscilloscope display can be projected conveniently on the classroom screen. With minimal investment in hardware (a microphone and a multimedia-ready PC), students may have a lab experience while in the classroom.

Some of my favorite apps in the Android platform include *OsciPrime*, *gStrings*, and *Sound Meter*. Each of these apps helps demonstrate the waveform, pitch, and loudness of sound. *OsciPrime* is a simulated oscilloscope that is capable to display the wave form of a sound source. The frequency, amplitude and wave form indicate the pitch, loudness, and timbre of a musical tone respectively (see Fig. 10.1).

Like a digital tuner, *gStrings* can tune the pitch of any given note of the chromatic scale. In addition, it can also identify a given pitch, provide the closest letter name of the pitch, and show how much deviation (in Hz) from that pitch. *Sound Meter* measures the loudness of a tone in dB and provides a time chart to indicate the change of volume. It also provides a reference chart of certain events and their sound volume; for example, the sound of a ticking watch or rustling leaves is about 20dB, a quiet whisper at three feet in a library is about 30dB, an alarm clock is about 80dB, while rock music or a screaming child is about 110dB! (See Fig. 10.2.) These tools, among others, have been helpful to visualize more abstract concepts of sound.

One of the harder musical concepts to teach in the introduction class is the monophonic versus polyphonic texture of music. On the one hand, polyphonic music of the fourteenth to eighteenth centuries is less familiar to students who have not been exposed to Western historic music; on the other hand, being able to identify aurally the numerous independent musical lines in polyphonic music for the untrained ears is not an easy task. Many students get confused with true polyphonic texture in music from music played by several instruments. Analyzing the music scores is a viable pedagogical approach, but the skills required to do so are generally beyond the ability of most students in that particular course. Then I came

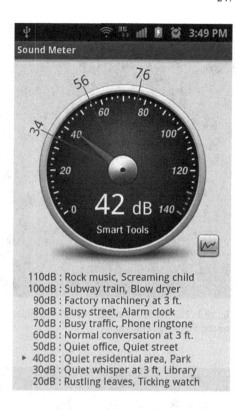

Fig. 10.2 A screenshot of the Sound Meter Android app

across with an incredibly visualized approach to music appreciation—The Music Animation Machine (www.musanim.com) by Stephen Malinowski. Essentially, the tool offers a visual rendition of a piece of music, with moving color bars that indicate the pitch and the duration of the musical tones. In Malinowski's words, the "different colors denote different instruments or voices, thematic material, or tonality. And each note lights up at the exact moment it sounds, so you can't lose your place." This is a marvelous product. It not only enhances the listening experience with visual effects, but also provides an excellent pedagogical tool to show the music parts when they are performed. I often show the Music Animation Machine rendition of J. S. Bach's *Toccata and Fugue in D minor* (http://www.youtube.com/watch?v=ipzR9bhei_o) as a demonstration of homophonic and polyphonic music (See Figs. 10.3 and 10.4).

As mentioned before, everyone learns differently; some are auditory learners, while others are visual or kinesthetic learners. Music is the art of sound in time, therefore it has natural tendency to favor the auditory approach to teaching and learning. While this may be effective in many circumstances, it does have its limitation when a learner, particularly a non-practitioner of music, who cannot grasp the different qualities of different sound. Computer animation as simple as the waveform of sound, to the more sophisticated rendition of musical themes and texture by the Music Animation Machine, has provided an invaluable tool to the teaching and learning of music.

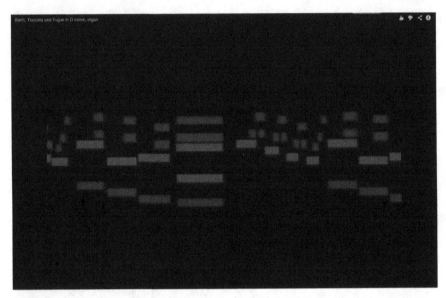

Fig. 10.3 The Music Animation Machine rendition of a homophonic passage of J. S. Bach's Toccata and Fugue in D minor

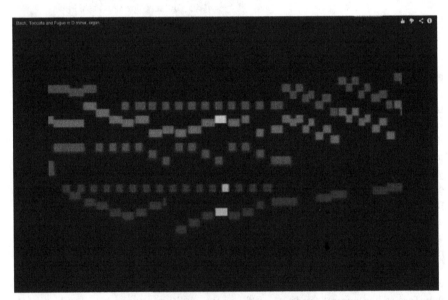

Fig. 10.4 The Music Animation Machine rendition of a polyphonic passage of J. S. Bach's Toccata and Fugue in D minor

10.3.4 General Technological Usage

10.3.4.1 MIDI and Keyboard

The Musical Instrument Digital Interface, or MIDI for short, is a protocol that allows electronic musical instruments and computers to connect and communicate with each other. Electronic musical devices, such as the early synthesizers, became increasing common in the 1970s. Instruments developed by different manufacturers, however, were incompatible with each other as these manufacturers employed different communication protocols among the products. The technical standards of MIDI, developed by the leading manufacturers of electronic musical instruments, were published in 1982, and gradually became industry standards for both the pop music and classical music arena. Among the many types of MIDI controllers, the most common one is the piano-style keyboard. A modern day synthesizer combines a piano-style keyboard (which acts as the controller) and an integrated electronic device that can generate a variety of sounds. These sounds often imitate other musical instruments. Some are capable to generate other tones such as a whistle or a wailing sound.

In terms of music pedagogy, MIDI and the synthesizer have opened new opportunities in the classroom. The portability and affordability of electronic keyboard (or as simple as a digital piano) provides accessibility to almost maintenance-free musical instruments in the classroom and the practice studio. Some schools have piano labs that feature all electronic keyboards or digital pianos. As mentioned earlier, these keyboards are very helpful for the students' individual learning as well as group learning.

Computer-connected MIDI instruments provide yet other opportunities for teaching and learning. Through the use of sequencing software, the computer allows a composer or arranger to cut, copy, or paste segments of music. Transposition becomes a lot easier both on notation and on performing.

10.3.4.2 Notation

The computerization of musical notation is definitely a welcomed progress for musicians, teachers, and students. A computerized notation system depends on the graphic display capability of the computer, as the character-based display is not adequate to present musical staff, notes and other symbols on the monitor screen. The development of the VGA standards in 1987 has made computer notation possible.

As mentioned earlier, most modern day computer notation software have eliminated the tedious processes of copying music by hand, from the orchestra scores to the individual parts. Making small changes to a newly composed or arranged piece no longer require the preparation of manual manuscripts for the entire piece or the entire page. Editing and transposition are less time consuming than before.

Modern technology allows USB ("Universal Serial Bus") connection of a MIDI keyboard to a computer. This has greatly facilitated the notation process. Instead of using the typewriter-style keyboard or the mouse, a student may use a piano-style keyboard to play a single melodic line or multiple note chords. The computer captures all the notes being played, and displays them properly on the musical staff. Computer based notation software has become an indispensible tools for musicians and teachers.

10.3.4.3 Communication and Distance Education

The gain in popularity and accessibility of the Internet has opened new and viable options for music education in the last decade. Thanks to the increasing bandwidth available to many households, and the lowering of hardware costs every year, many households now have broadband Internet connection. Even remote areas are served by increasingly affordable satellite-linked Internet connection. Webster (2007) enthusiastically anticipated that the constant advancement of technology, particularly in the audio and video capability over the Internet, would increase the effectiveness of the delivery of distance music education.

Indeed, we have seen more and more music educational programs delivered effectively over the Internet since Webster wrote his essay several years ago. Recently, I interviewed Dennis Chan, one of my college classmates who is now operating a very successful music school both online and "on ground." ("On ground" is a buzzword for the distance education community referring to physical classroom, or face-to-face instruction.) By using the web conferencing environment and the small group setting, the computer screen becomes an extension of his classroom and studio. As a pioneer in the area of Internet delivery of music education, particularly in the field of music theory, he has invested a lot of time and energy in making all of his presentations and teaching "project-able"—meaning that he has eliminated totally the need of the traditional chalkboard, whiteboard, or even the "Smart Board."

Like webinars, Internet-based music education enjoys the advantages of reaching many more people when the physical distance between the learners and the teacher is no longer a hurdle. Current technology can not only afford a one-to-many communication (such as one teacher talking to a group of students) or vice versa (such as a group of students can talk to the teacher or moderator), but also provide a platform for instant peer communication. Here, instant peer communication refers to the capability that students attending the virtual classroom can simultaneously interact with each other. For example, when one of the students raises a question, other students may interject and offers their insight on the question. Dennis asserts that while the presence of a moderator in a webinar setting may be helpful, it is less crucial when the class is in a small group setting.

Very interestingly, Dennis not only offers online music theory lessons, but also violin lessons over the Internet. His success in online music lessons may again attribute to the increased capability of the computer hardware (such as the fidelity of audio and video quality), the increased availability and affordability of broadband Internet connection at home, and the increased acceptance of the concept of online classroom and studio. Ten years ago, the latency of three seconds in video communication was a major hurdle if the teacher desired to play together with the student. The delay, in terms of either the video or the audio element, would make a duet between the teacher and the student impossible. One could imagine that even a slight delay would bring a high level of frustration for the experience, let alone a three-second delay. The current broadband speed, together with the video processing capability of the more advanced video systems available on microcomputers, make online studio teaching a viable alternative to the physical presence.

Admittedly, there are many challenges in online music education. Firstly, in the previous example of giving online violin lessons, the teacher cannot have physical interaction with the student in an online environment. This limitation would make it impossible if the teacher wants to correct the student's fingering, his/her touch on the string, or correct the student's posture. An online environment seems to be more suitable for more advanced players when each class is structured as a masterclass. A well-prepared student would make a lesson more effective and efficient. Secondly, online learning, particularly online music education, seems to appeal to people of certain age groups. Undoubtedly, as the younger generations grow up with more exposure to advanced technology, they are more technologically savvy and more open to the idea of taking their education online. Today, many young people have access to multiple devices that have WiFi or data capability over a cell phone network (for example, a computer, a tablet, a cell phone, an e-reader, etc.). The accessibility and availability of such telecommunication devices makes distance learning more attractive to young people. Thirdly, a certain size of the population casts doubt on the effectiveness and quality of online music education. As my music teacher friend points out, the people who raise the most concern about online music education are the parents. On the one hand, many baby-boomer parents did not experience online education during their school years (although some may have been exposed to TV education); on the other hand, baby-boomer parents in general are quite involved in the decisions of their children's education. Their lack of personal experience, and sometimes trust, in online education could become a hurdle for the younger generation to take part in distance music education.

That being said, the future of online music education seems promising, however. The contributing factors for its future growth include (a) the continuing upgrade of hardware, (b) the steady decrease in price of computers and other devices, (c) the improvement of the Internet backbone and infrastructure, and (d) quite importantly, the increase in acceptance and readiness for the music teachers.

10.4   Fuzzy Logic and Artificial Intelligence

Music is an art as well as a science; therefore the teaching of music should address both the art and the science of music making. As we can see in the foregoing pages, the computer plays an important role in music pedagogy in the twenty-first century. Since the computer is a precise instrument, when it is used in teaching such as the question-and-answer drills, it looks for a precise answer that either matches or deviates from the predetermined value. In this regard, the computer is an invaluable pedagogical tool for the right-or-wrong approach. Yet music, particularly in performance practice, is more of an art than a science. Human voice, for example, encompasses a range of frequencies rather than a single and sustained absolute pitch. The same flexibility is also true for rhythm. No reasonable musician would expect an absolutely precise tempo that is expressed in milliseconds. A stated tempo such as *andante* allows for a range of speed, generally from 72 to 78 beats per minute (BPM). Even if we lock in to a specific tempo, say, 72 BPM, the interpretation of music such as phrasing would allow variations to the basic pulse. The loudness of music tones may have even larger room for interpretation. All of these factors point out a fact that certain usage in computer-aided instruction (CAI) in music needs flexibility within a reasonable range.

While the research in fuzzy logic and artificial intelligence in music is not new, it is an area that needs ongoing research and development. Fuzzy logic, developed in the mid-1960s and the early 1970s, deals with reasoning and "degrees of truth" rather than the usual Boolean logic of true or false. A detail discussion of fuzzy logic is beyond the scope of this essay; but it is suffice to say that its concept is comparable to a group of rules in the linguistic environment. As soon as several rules are in place, fuzzy models begin to function.

For a computer application to be implemented successfully in the ear-training CAI environment, for example, the decision processes of the computer involve the determination if the response is acceptable. The response is sometimes in a different octave (such as the natural differences in the man's and woman's voice), or the range of fluctuation in terms of pitch and timing. The music education software industry seems to be steadily moving to the direction of developing software that is more "intelligent" in this regard. EarMaster ApS, a Danish music software company, develops and markets a product called EarMaster (www.earmaster.com). Now in version 6, the software is capable for a student to sing back a single melodic line or a part in a multi-part score, and receive instant feedback on the accuracy. Similar exercises are available rhythmic patterns.

Artificial intelligence (AI), a term coined by John McCarthy in 1955, is defined as the science and engineering of making intelligent machines. It involves a system that perceives its environment and takes actions that maximize its chances of success. AI, though not as popular in music as in other fields, has opened new opportunities in music education and application. Some of the key players in this field include Sinfonia and SmartMusic, among others. Sinfonia is a program designed to provide orchestra enhancement; it is capable to follow a conductor's tempo and respond to constantly changing music nuance during live performance (www.rms.biz/products/sinfonia/about). Smart Music is an interactive practice tool that can detect mistakes and react to the student's tempo changes.

10.5 Summary

In the past few decades, the computer, whether in its general usage and specialized applications, is making noticeable impact in music pedagogy. In general usage, the computer enhances the presentation and delivery of the music teaching experience.

The general increase in computer literacy among the newer generations of teachers and students has made the computer an indispensable tool in teaching and learning. The enhanced accessibility and speed of the Internet now afford distance learning, from the delivery of lecture to studio teaching of a musical instrument. Music enthusiasts of the millennial generation can even learn musical techniques through YouTube. In specialized music applications such as notation, MIDI sequencing, ear-training, sight reading, and accompaniment, the computer has expanded the where, when and how these techniques are taught and practiced. As technology keeps moving forward, it will be exciting to see more powerful and human-like applications in the years to come. At that time, the computer could turn a new page in music pedagogy.

References

Berz WL, Bowman J (1994) Applications of research in music technology: from research to the music classroom. R & L Education, Lanham

Higgins W (2002) Technology. In: Colwell R (ed) Handbook of research on music teaching and learning. Schirmer Books, New York

Kuhn WE, Allvin RL (1967) Computer-assisted teaching: new approach to research in music. J Res Music Educ 15: 305–315

Ruthmann A (2006) Negotiating learning and teaching in a music technology lab: curricular, pedagogical, and ecological issues. Dissertation, Oakland University, pp 21–27

Webster PR (2002) Computer-based technology and music teaching and learning. In: Colwell R, Richardson C (ed) The new handbook of research on music teaching and learning: a project of the music educators national conference. Oxford University Press, New York, p 416 ff

Webster PR (2007) Computer-based technology and music teaching and learning: 2000–2005. Springer Int Handb Res Arts Educ 16:1311–1330

Index

2D-FMC *See* 2-dimensional Fourier
  Magnitude Coefficients, 179
2-dimensional Fourier Magnitude
  Coefficients, 179
50 Cent, 8
101 Dalmatians, 14
2001:
  A Space Odyssey, 16
[xor] synth, 202
β rhythm, 211
β-wave, 211, 212

A

A2IM *See* American Association of
  Independent Music, 6
Ableton Live, 131, 132
Ableton Push, 132
absolutist view, 137
acoustic fingerprint, 4
acoustic instrument, 52
acoustic signal, 102
Advanced Audio Coding, 4
Advancing Interdisciplinary Research in
  Singing, 72
aesthetic attitude, 139
aesthetic emotion, 139
aesthetic experience, 139, 140, 145
aesthetic judgment, 139
aesthetic perception, 128, 146, 157
aesthetic preference, 139
aesthetics of sonification, 189
aesthetic stimuli, 139
affect, 48, 68
affectation, 30, 48, 49, 50, 51, 52, 53, 54, 55,
  56, 57, 64, 65, 66, 68, 69, 72
affective expression, 89
affective sound, 186
Ahmet Ertegun Award, 1
Air-Quality Egg, 185

Akkersdijk, S., 86, 88
Albin, Zak J., 40
Albon, S. D., 61
aleatoric music, 120
aleatoric procedures, 37, 131
algorithmic composer, 138
algorithmic composition, 38, 132, 133
algorithmic composition engine, 91
algorithmic improvisation system, 101
algorithms, 31
al-Jazarī, 36
Allvin, Raynold L., 240
Almighty, The, 21
alpha band, 206
alpha leader, 60
alto clef, 241
Amazon mp3, 17
ambient music, 129, 130
ambient sound, 130
American Association of Independent
  Music, 6
American Idol, 9, 25
amplification, 53
amplifiers, 51, 52
amplitude envelope, 131
analysis of variance, 210
andante, 252
Andrade, Mark, 14
Andrews, Kevin, 21
Angel of Death, 138
animal communication, 46, 60
animal learning, 140
animation, 247
annotation, 87
ANOVA *See* analysis of variance, 210
Antares, 49
anti-apartheid anthem, 19
anticipatory paradigm, 137

Printed in the United States
By Bookmasters